# Gunn-effect Logic Devices

By the same author

*Semiconductor Plasma Instabilities*

# Gunn-effect Logic Devices

**Hans L. Hartnagel,** Dipl.Ing., Ph.D., D.Eng., M.I.E.E.,
Sen. M.I.E.E.E., F.I.E.R.E.
*Professor of Electronic Engineering,*
*University of Newcastle upon Tyne*

**American Elsevier Publishing Company, Inc.**
**New York**

American Edition published by

American Elsevier Publishing Company, Inc.
52 Vanderbilt Avenue
New York, New York, 10017

Library of Congress Catalog Card No: 73–14122
ISBN–0–444–19548–3
© Hans L. Hartnagel 1973
First published 1973

Published by
Heinemann Educational Books Ltd
48 Charles Street, London W1X 8AH

# Preface

It is an accepted practice in modern scientific and technological developments to publish new results first in professional journals. After a while this material has to be reviewed and put into correct perspective. It is therefore useful to discuss and evaluate the achievements in a book.

Gunn-effect logic was proposed very shortly after the publication of Gunn's discovery. In fact, there are references to this possibility in the text of his patent, which is held by the large computer company IBM. Many developments have taken place so far and impressive device results have now been achieved in several countries. There are plans to consider Gunn-effect logic for the regeneration of gigapulse signals in future wide-band PCM communication systems. It is therefore the correct moment to present an overall assessment of this new field.

As soon as a new component becomes a technological reality, there is a demand for an introductory book for the many engineers and scientists who will in some way or other become involved with the subject. This is now also the case with Gunn-effect pulse-processing devices. This book tries to satisfy this need.

Finally, many people would surely like to inform themselves about new developments in the fields of ultrafast pulse communication and signal processing. This book should be of help here also. It is stressed, however, that a general knowledge of mathematics and semiconductor physics at the level of undergraduate courses in, say, electronics or physics, is necessary for full comprehension of the aspects covered.

The author has been involved with the development of this subject from its beginning, and he has kept in close contact with the majority of the laboratories where contributions to Gunn-effect logic are made. It is a good opportunity now to express his gratitude to all the specialists in this subject for many stimulating discussions and the free exchange of ideas and experience. It is difficult to single out any one by name without committing an injustice. However, he would like to thank especially Professor H. Yanai of the University of Tokyo, Japan, who also made some of the figures available from his own work. Similarly, he is thankful to those who worked with him in the past on Gunn-effect logic, namely T. Izadpanah, W. Fallmann, M. Kawashima and several others. Of course, he must not forget to mention the patience and active help of his wife regarding the preparation of this manuscript.

As an author of a book always knows his own work best, his own results might obtain unjustifiable preference over those of other workers in the field. The present author hopes that he has achieved a reasonable balance of the details described. Sometimes his own data is used in order to give a particular example out of the many ways in which an approach can be made to a problem.

It is hoped that this book will be found useful by the reader, and that it helps further to develop the art of Gunn-effect logic.

1973                                                                    H. L. H.

# Contents

# Nomenclature

$A$ = cross-sectional area of current flow

$A_E$ = amplitude of electric field component

$A_S$ = spectral amplitude

$A_T$ = cross-sectional surface of heat flow through substrate

$A_\mu = 1 + \dfrac{\mu_0}{\mu_n}$

$B$ = magnetic induction

$B_E$ = amplitude of electric field component

$B_\mu = \left[ \dfrac{\mu_0}{\mu_n} + \left( \dfrac{\mu_0}{\mu_n} \right)^2 \right]^{\frac{1}{2}}$

$C$ = capacitor

$C_d$ = domain capacitance

$D$ = diffusion constant

$D_G$ = a diode

$D_1 D_2$ = Gunn-effect diodes

$E$ = electric field

$E_M$ = peak field

$E_R$ = field outside the domain, in connection with mature domain, also called rest field

$E_{R0}$ = field outside the domain for zero domain amplitude

$E_{R\infty}$ = field outside the domain for fully grown domain

$E_a$ = average field = $V_B/l$

$E_d$ = domain field

$E_e$ = domain extinction field

$E_g$ = critical field under gate

$E_0$ = material constant with the units of field; or: d.c. field component

$E_p$ = operating field

$E_s$ = electric field for three-line approximation of $v(E)$ where satellite velocity is reached

$E_t$ = threshold field

$E_u$ = uniform field for case where $I = I_0$ due to $n = n_0$

$E_1, E_2, E_{ir}$ = field values

$F$ = a function

$H$ = magnetic field

$H_0$ = magnetic d.c. field

$I$ = total current

$I_D$ = diode current during domain transit

$I_n$ = integral

$I_0$ = total current for cases where $I = I_0$ due to $n = n_0$

$I_t$ = threshold current for domain nucleation
$J$ = current density
$J_{\text{diff}}$ = diffusion current
$J_v$ = valley current
$L$ = inductance
$L_D$ = Debye length
$L_L$ = equivalent loading inductance

$$L_s = L_D \left(\frac{\mu_0}{\mu_n}\right)^{\frac{1}{2}}$$

$$M = \frac{V_A}{V_t}$$

$M_i$ = number of valleys of type $i$
$N_i$ = density of states of valley of type $i$
$P = V_t^2/R_0$, power
$P_L$ = power delivered to the load
$P_S$ = signal power
$P_g$ = power generated by device
$P_i$ = input power
$P_{im}$ = input power triggering domains reliably
$P_l$ = losses of cavity
$P_0$ = output power
P', P" = operating points
$P_1, P_2, P_{1r}$ = potentials
$Q$ = charge stored in accumulation layer per unit area
$Q_e$ = external quality factor of cavity
$Q_l$ = loaded quality factor of cavity
$Q_m$ = maximum domain charge
$Q_t = \varepsilon E_t$
$Q_0 = Q/\exp(\omega_{dn}t)$
$Q_1$ = charge at time $t_1$
$R_L$ = load resistance
$R_{LP}$ = resistance in parallel with load
$R_0$ = low-field resistance of Gunn-effect devices
$R_1, R_2$ = resistances
$S_p$ = synchronization pulse
$S, S_a, S_b$ = summation
$S_1, S_2$ = imput signals
$T$ = temperature
$T_L$ = time constant of loading case
$T_a$ = growth time for $E_M > E_s$
$T_b$ = growth time for $E_t < E_M < E_s$
$T_e$ = electron temperature
$T_g$ = growth time of domain
$T_1$ = temperature of active layer

$T_0$ = lattice temperature
$T_r$ = repetition time of pulses
$T_s$ = temperature at substrate-metal interface
$T_\gamma = \dfrac{2}{\gamma \, \Delta H}$
$T_1 = L/R$
$T_2$ = time constant
$V_A$ = bias voltage threshold for ionization
$V_B$ = bias voltage
$V_{Bm}$ = minimum bias voltage
$V_D$ = excess voltage absorbed by Gunn-effect domain
$V_{Dm}$ = excess domain voltage for full maturity
$V_{Dth}$ = excess domain voltage producing 1 per cent of $J_v$ increase
$V_{D0} = V_D/\exp \omega_{dn} t$
$V_E$ = extinction voltage for oscillations
$V_S$ = voltage amplitude of retiming spike
Vol. = volume of active device
$V_s$ = signal voltage
$V_t$ = threshold voltage = $E_t l$
$W$ = energy stored in cavity
$X$ = second lowest energy valley of GaAs
$Z_B$ = bias line impedance
$Z_L$ = load impedance
$Z_i$ = input line impedance
$a$ = half thickness of active layer
$a_1$ = width of loading layer
$a_n$ = constants of a series
$a_0, a_1, a_2$ = constants of a series
$c$ = a 'constant', see equation (2.61)
$c_n$ = value of $c$ for increment $T_n$
$c_0$ = velocity of light
$c_I, c_{II}$ = light velocity in materials I and II respectively
$d$ = thickness of $n$-layer = $2a$
$d_a$ = thickness of accumulation layer
$d_d$ = width of domain depletion layer
$d_s$ = thickness of semi-insulating substrate
$e$ = electronic charge
$f$ = frequency
$f_b$ = base-band repetition frequency
$f_h$ = frequency of high pulse-repetition rate
$f_1$ = frequency of low pulse-repetition rate
$f_r$ = repetition frequency of spikes
$f_t$ = frequency of trigger signal
$f_0$ = free running frequency
$f_1$ = frequency representing 'mark'

$f_2 =$ frequency representing 'space'

$g = \frac{1}{2}(1 - k)\mu_0/\mu_2$

$g_d =$ r.f. conductance in microwave device

$g_1 =$ load conductance

$h_g =$ ratio of field increase for gate voltage increase of FET

$k =$ inverse of peak-to-valley current ratio

$k' = 1 - k$

$k_B =$ Boltzmann constant

$k_w = \omega/c$

$l =$ interelectrode distance of Gunn-effect device

$l_1 =$ effective device length which includes $R_L$

$l_t =$ distance from gate to anode

$m =$ integer

$m_f =$ field ratio

$m_i =$ effective electron mass for valley type $i$

$n =$ electron density

$n =$ integer

$n_a =$ density of accumulation layer

$n_c =$ coupling coefficient $= v_{01}/v_{ac}$

$n_d =$ minimum electron density of depletion layer

$n_i =$ number of electrons in valley of type $i$

$nl_{st} = nl$ product for subthreshold biasing

$n_0 =$ donor density

$q =$ number of pulses

$s = n + 1$

$t =$ time

$t_d =$ delay time

$t_r =$ rise time

$t_t =$ domain transit time

$t_1 =$ time when field reaches $E_s$

$u(t) =$ velocity of domain

$v =$ electron velocity

$v_D =$ domain velocity

$v_R = v(E_R)$

$v_{ac} =$ amplitude of r.f. voltage across device

$v_d =$ d.c. velocity

$v_r =$ reflected signal

$v_s =$ saturation velocity for high fields

$v_t =$ threshold velocity for transferred-electron effect

$v_0 = \mu E_0$

$v_{01} =$ amplitude of r.f. voltage across the load

$w =$ transverse width of interface or Gunn-effect diode

$x = y_d/d$

$x_1 =$ distance

$y_d$ = depth of depletion layer under gate

$y_i$ = distance

$z_a', z_b'$ = two points of the domain

$z_a$ = point of domain where $n = n_a$

$z_\alpha$ = point of domain where $n = n_\alpha$

$z_q$ = moving point with no charge crossing during domain growth or decay

$z_1, z_2$ = two points on either side of the domain

$\Gamma$ = lowest energy valley of GaAs

$\Delta$ = energy difference between lowest energy valley to second lowest valley

$\nabla$ = nabla operator

$\Delta E$ = reduction in bias field

$\Delta I_D$ = current drop due to domain nucleation

$\Delta T, \Delta T_a, \Delta T_b$ = growth time increment

$\Delta f$ = frequency shift

$\theta$ = angle

$\theta_T$ = thermal resistance

$$\theta_e = \frac{kT_e}{\Delta}$$

$$\theta_0 = \frac{kT_0}{\Delta}$$

$\Phi_d$ = pinch-off voltage

$\alpha$ = field decay constant, from $E_y = A_E \sin \alpha y$

$\alpha_F$ = attenuation constant of circularly polarized wave in ferrimagnetic materials

$\beta$ = propagation constant

$\beta_F$ = phase constant of circularly polarized wave

$$\beta_{cy} = \frac{\mu_y n_0 e}{\varepsilon v_0}$$

$$\beta_{cz} = \frac{\mu_z n_0 e}{\varepsilon v_0}$$

$$\beta_e = \frac{\omega}{v_0}$$

$\gamma$ = figure of gate trigger capability

$\varepsilon$ = permittivity

$\varepsilon_{ci}$ = conduction valley energy of type $i$

$\varepsilon_f$ = permittivity of free space

$\varepsilon_r$ = relative permittivity

$\varepsilon_0$ = Fermi level

$\kappa$ = thermal conductivity

$\kappa_0, \kappa_1$ = constants of thermal conductivity

$\mu_d = \mu_y$ d.c. mobility

$\mu_i$ = mobility of electron in valley of type $i$
$\mu_n$ = negative differential mobility
$\mu_r$ = relative permeability
$\mu_y$ = mobility in transverse direction
$\mu_z$ = mobility in longitudinal direction
$\mu_0$ = low-field mobility
$\rho$ = low-field resistivity
$\tau$ = pulse duration
$\tau_d$ = dielectric relaxation time
$\tau_e$ = energy relaxation time
$\tau_m$ = minimum pulse duration required for domain nucleation
$\tau_n$ = dielectric relaxation of the modulus of the negative differential resistivity
$\omega_d$ = dielectric relaxation frequency
$\omega_{dn}$ = $en_0\mu_n/\varepsilon$, dielectric relaxation frequency for negative mobility
$\omega_{do}$ = $en_0\mu_0/\varepsilon$
$\omega_m$ = $\gamma 4\pi M_0$
$\omega_0$ = angular frequency of operation
$\omega_t$ = resonance angular frequency

# 1:   Introduction to Gunn-effect Logic

Many applications of pulse-processing electronic circuits became possible with the advent of semiconductor junction devices. A further impressive expansion can be foreseen with the possibilities of cheap and reliable large-scale integration. These junction devices are not only bipolar, such as junction diodes and bipolar transistors, but also various types of field-effect transistors, and here we have to include the MOS transistor, which of course, strictly speaking does not rely on any junction.

However, when increased pulse rates are demanded, various difficulties arise. The junction or MOS capacitance has to be charged or discharged via the transmission line of some characteristic impedance which is commonly 50 $\Omega$, and which should not be further reduced by any appreciable amount as the losses would otherwise become too large or the geometry of the line unrealistic. The corresponding capacitance-impedance product is then typically around a few hundred picoseconds and can possibly be reduced to something like 50 ps. During these short times the input impedance of junction devices is, however, more complex than can be represented by a simple junction capacitance so that the input characteristics are highly dispersive. Additionally, for bipolar transistors the real part of the input impedance at microwave frequencies is low and has to be matched to the characteristic impedance of the transmission line via an impedance transformer. Such a transformer is, however, again highly dispersive, unless large losses are acceptable.

The frequency spectrum of a pulse wave shape is shown in Figure 1.1, and the time function can only be maintained if the transmission system leaves the general shape of the frequency spectrum unaltered. A dispersive network such as is given by bipolar transistors at microwave signals can, however, affect the time function by such an amount that spurious pulses might appear as sketched in Figure 1.2, and the information content is seriously distorted. When spectral distortions are not too strong, applications where high repetition rates but not large processing speeds are required, so that delays can be accepted, permit the use of compensation networks which show the inverse frequency characteristic from that of the dispersive device. In such cases impedance transformers can also be accepted. These applications are given where, for example in gigapulse-rate communication systems, pulses are regenerated by individual active devices. Sometimes the dispersive characteristics of junction devices are, however, such that parts of the spectrum can be lost below the general noise level with the result that compensating networks cannot be employed any more either. Therefore it is now generally accepted that the limit for junction devices lies somewhere at 1 gigapulse/second. This does not mean,

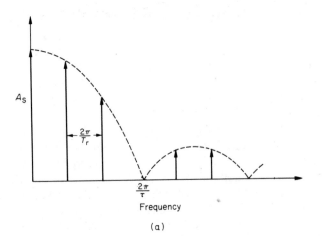

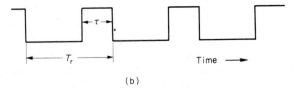

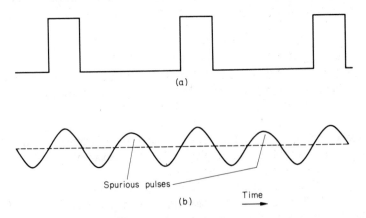

**Figure 1.1** (a) is the spectrum of pulse wave (b) with $T_r$ the pulse repetition time and $\tau$ the pulse duration.

**Figure 1.2** (a) original signal, and (b) its low-frequency distorted signal which exhibits false pulse signals.

of course, that such devices can be useful for microwave applications where a limited bandwidth is required.

Bipolar junctions have the additional disadvantage of minority carriers, which have to be removed from the junction when the device is switched from the forward to the reverse direction. These minority carriers cause a reverse current to flow during the first part of the reverse-bias period until the junction charge is depleted. Such difficulties are not encountered with metal-semiconductor junctions where only one type of carrier is involved (for a discussion of Schottky diodes see reference 29). Therefore Schottky-gate FETs have been found to operate very fast, as long as the gate width is kept sufficiently short. The gate width has to be reduced because the switching time is given by the time it takes the carriers to travel underneath the gate. Therefore high saturation velocities are advantageous, and GaAs Schottky-gate FETs have produced the fastest performance so far. The gate is then only 1 $\mu$m wide and advanced photolithography is required.

When pulse rates above 1 gigapulse/second have to be handled, other pulse-processing effects have to be utilized if these difficulties are to be avoided. Therefore, the transferred-electron effect[6,20] has been considered, because high-field domains can produce here very fast current switching. A Gunn-effect device has a very small capacitance before a domain has been nucleated. This capacitance is given by the electrode separation of a volume device and is therefore at least two orders of magnitude smaller than with junction devices. A very short input pulse can nucleate a domain which then grows independently to a mature size, and the device current is switched to a low-current state. As one has a threshold device here, the value of the required input-signal amplitude depends on the stability of the bias voltage. With cheap commercial I.C. units which are presently available for voltage stabilization, this requirement of bias stability is easily met. The threshold property of this effect can then be exploited for logic applications. Additionally, the current reduction will only continue until the Gunn-effect domain reaches the anode electrode and the current then reverts back to its original level. One has therefore a monostable element that can be employed easily for the reshaping of pulse signals. There is also a possibility of exploiting a stationary domain under certain conditions,[60a,b] and one can obtain a bistable pulse device. This would be particularly useful for fast memory applications.

Gunn-effect pulse devices as simple two-terminal elements have no directionality property, and difficulties can arise with reflections and instabilities. It is therefore of advantage to have three-terminal elements in the same way as transistors have three terminals. There is firstly the possibility of depositing a Schottky-gate electrode in front of the cathode. This can constrict the current path, resulting in a local high field. Domain nucleation can then easily be achieved by applying a pulse signal to this electrode so that the space charge layer is further extended and the fields there are increased above the domain-forming threshold. One has then, of course, a junction capacitance again and the ultimate pulse rates achievable are limited. However the transferred-electron instability aids switching which is then still reasonably fast. Additionally, the many other advantages of these domain devices still exist, such as high-current switching with reasonable impedances, and the reshaping facility of pulses without having to use some flip-flop circuitry which would introduce more complexity. There is a good chance that purely resistive three-electrode Gunn-elements can be developed for pulse processing, and then the speed disadvantages of Schottky trigger electrodes would be avoided.

Although Gunn-effect pulse-processing possibilities were realized a long time ago, namely shortly after Gunn's discovery of travelling high-field domains, progress has been slow so far. This has partly been caused by the lack of good GaAs materials, and partly by a certain doubt regarding reliability, packaging density limits, and the importance of going to gigapulse rates. However, steady progress has been made over these last years, and numerous applications have been envisaged now that d.c. biased devices have been developed and operated successfully by several laboratories. Once the first application of Gunn-effect pulse devices has been realized, other applications will follow soon. Therefore it is important to introduce this new field to a wider range of electronic engineers and physicists than those who have been concerned with this subject hitherto.

It is assumed that the reader has a good knowledge of semiconductor electronics, whereas no familiarity with the transferred-electron effect is required. This book aims, therefore, at electronic engineers and physicists who possess an understanding of semiconductors corresponding to that of an undergraduate reading electronics at a university in his second or third year.

# 2:  Gunn-effect Domain Dynamics

In this chapter the basic principles of Gunn-effect domains are reviewed in as far as is required for the subject of this book. The space devoted to this purpose is kept to a minimum, as the subject of the transferred-electron effect has already been described by other authors. If the reader is interested in any detailed description of the various transferred-electron oscillators or the theory of electron transport in GaAs, for example, he is referred to these books.[6,20]

## 2.1  BASIC PRINCIPLES OF A BIPOLAR DOMAIN

When the energies of conduction electrons in semiconductors are computed as a function of their momenta, a range of valleys of the energy contours is obtained. Of course, practically all conduction electrons are positioned in the lowest valley unless they are heated up. A convenient way of heating charge carriers is the application of a high electric field. If we take $\Delta$ to be the energy difference between the bottom of the lowest valley, $\Gamma$, and the bottom of the second-lowest valley, X, the electrons have to acquire a temperature $T = \Delta/k_B$, before they can be transferred from $\Gamma$ to X.

If $N_i$, the density of states for each valley, is known, one can write down the number of electrons $n_i$ in each valley of type $i$ (see section 1.3 of reference 20):

$$n_i = M_i N_i \exp\left[-(\varepsilon_{ci} - \varepsilon_0)/k_B T_e\right]$$

where $M_i$ is the number of valleys of type $i$, and $T_e$ is the electron temperature. Using the values of the effective masses $m_i$ one can say

$$\frac{N_1}{N_2} = \left(\frac{m_1}{m_2}\right)^{\frac{3}{2}}$$

The resulting number ratio of electrons is then

$$\frac{n_2}{n_1} = \frac{M_1}{M_2}\left(\frac{m_1}{m_2}\right)^{\frac{3}{2}} \exp\left(-\Delta/k_B T_e\right) \tag{2.1}$$

The temperature of the electrons can be estimated with the help of the concept of energy relaxation time $\tau_e$, i.e.

$$eEv = \frac{\frac{3}{2}k_B(T_e - T_0)}{\tau_e} \tag{2.2}$$

where $v$ is the average electron velocity over both valleys, i.e.

$$v = \frac{n_1\mu_1 + n_2\mu_2}{n_1 + n_2} E \tag{2.3}$$

and $T_0$ is the lattice temperature. We do not assume here for our simple calculation that the mobility is field dependent for each valley, so that our results are incorrect for higher field values where saturation sets in. As the satellite velocities are very small, we set $\mu_1 \cong 0$. Using equations (2.1) and (2.3) we then obtain

$$\frac{v}{v_o} = \frac{E}{E_0}\left[1 + \frac{M_1}{M_2}\left(\frac{m_1}{m_2}\right)^{\frac{3}{2}}\exp\left(-\frac{1}{\theta_e}\right)\right]^{-1} \tag{2.4}$$

With $v_0 = \mu_1 E_0$,

$$E_0 = (\tfrac{3}{2}\Delta/e\mu_1\tau_e)^{\frac{1}{2}}$$

$$\theta_e = \frac{k_B T_e}{\Delta}, \qquad \theta_0 = \frac{k_B T_0}{\Delta}$$

where $T_e$ is related to $E$ via equation (2.2), i.e.

$$\theta_e = \theta_0 + \frac{E}{E_0}\cdot\frac{v}{v_0}$$

A numerical evaluation is shown in Figure 2.1. The lattice temperature of 290 K corresponds to $\theta_0 \sim 0.07$.

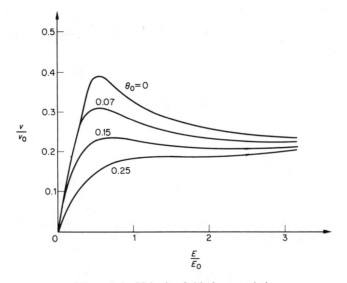

**Figure 2.1**   Velocity-field characteristics.

From Figure 2.1 one sees that there is a well defined threshold for the onset of the negative differential mobility. This negative mobility disappears for increased lattice temperatures (increasing $\theta_0$).

A disadvantage of the theory just outlined is the assumption of equal electronic temperatures in both types of valleys. This is incorrect. The satellite valley electrons gain energy from the electric field $E$, i.e. at the rate $en_2v_2E$. Unfortunately $\mu_2$ is very low so that this heating effect is weak. Additionally the hot electrons from the lower valley spill over into the upper valley and bring with them an electron temperature which is reduced by an amount corresponding to the potential energy $\Delta$. Moreover, the momentum exchange collisions with phonons required for the transfer to the satellite valley involves also energy exchange processes, affecting the temperature again. Additionally the energy losses in the upper valleys, caused by intervalley scattering, are more serious than for the central valley. Therefore it can be concluded that the upper valley electrons remain close to the lattice temperature except for very high fields, whereas only the lower valley electrons will be hot.

The intervalley scattering is the slowest process, and therefore limits the speed of response of the transferred-electron effect, because it is responsible for heating the central valley electrons sufficiently for them to be transferred to the satellite valleys.

To take all these scattering phenomena into account, a digital computer has to be employed for the determination of the velocity-field relation. Various experimental techniques have been employed for the measurement of this relation and good agreement with experiment has been achieved.

It is usually assumed that the $v(E)$ characteristic is not affected by the diffusion current $J_{\text{diff}} = -\nabla(eDn)$. In the two-valley transferred-electron model, the diffusion coefficient can be expected to be

$$D = \frac{k(T_1\mu_1 n_1 + T_2\mu_2 n_2)}{e(n_1 + n_2)}$$

As the electron temperatures and density ratios depend on the field, this indicates that $D$ depends on $E$ (see Figure 2.2). When there is then a diffusion current due to a density gradient, the energy balance is altered with the result that the electron temperature is changed. Therefore the above assumption is inconsistent unless diffusion is small. In fact, as some experimental results show a $D$ which is higher than that of the theory (see Figure 2.2), it seems doubtful whether this assumption is fully justified. However, one can obtain reasonable agreement between various experimental and physical data when diffusion is included by neglecting its influence on $v(E)$, and such practice should therefore be acceptable.

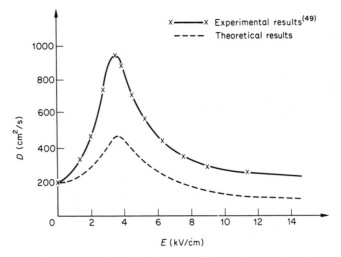

**Figure 2.2** The diffusion coefficient $D$ *vs.* electric field $E$.

The convection, diffusion and displacement currents respectively give the total current $I$ as follows:

$$\frac{I}{A} = env(E) - \frac{\partial(e\,Dn)}{\partial z} + \varepsilon\,\frac{\partial E}{\partial t} \qquad (2.5)$$

where $A$ is the cross-sectional area of current flow. Changes in field and charge density are related by Poisson's equation, i.e.

$$\varepsilon\,\frac{\partial E}{\partial z} = e(n - n_0) \qquad (2.6)$$

where $n_0$ is the donor density. For the semiconductor parts along $z$ where $n = n_0$ and the field $E = E_u$ is uniform, equation (2.5) becomes

$$\frac{I_0}{A} = en_0v(E_u) + \varepsilon\,\frac{\partial E_u}{\partial t}$$

$I_0$ describes the current at the places where the flow is not perturbed. As the total current must be continuous, we can set $I = I_0$. Additionally, we employ equation (2.6) and obtain

$$\varepsilon\,\frac{\partial}{\partial t}(E - E_u) = en_0[v(E_u) - v(E)] + \frac{\partial}{\partial z}(e\,Dn) - \varepsilon v(E)\frac{\partial E}{\partial z}$$

We integrate this equation over the space-charge perturbation from $z_1$ to $z_2$, where these points are just outside the perturbation so that the diffusion currents are equal there, and obtain

$$\frac{\partial}{\partial t} \int_{z_1}^{z_2} (E - E_u) \, dz = \frac{en_0}{\varepsilon} \int_{z_1}^{z_2} [v(E_u) - v(E)] \, dz - \int_{z_1}^{z_2} v(E) \frac{\partial E}{\partial z} \, dz \qquad (2.7)$$

The last term in this equation is also zero, i.e.

$$\int_{z_1}^{z_2} v(E) \frac{\partial E}{\partial z} \, dz = \int_{E_u}^{E_u} v(E) \, dE = 0$$

For small field changes when one can set

$$v(E) = (E - E_u) \frac{dv}{dE} + v(E_u)$$

the following equation is obtained from (2.7):

$$\frac{\partial}{\partial t} \int (E - E_u) \, dz = -\frac{en_0}{\varepsilon} \int (E - E_u) \frac{dv}{dE} \, dz$$

which yields

$$\frac{dV_D}{dt} = -\omega_d V_D \qquad (2.8)$$

with $V_D = \int (E - E_u) \, dz$ and the dielectric relaxation frequency

$$\omega_d = \frac{en_0}{\varepsilon} \frac{dv}{dE}$$

$V_D$ is the excess voltage as absorbed by the space charge perturbation. When $dv/dE$ becomes negative, $V_D$ will grow. From the results of Figure 2.1 it can be seen that a space-charge perturbation in a material exhibiting the transferred-electron effect can grow. This is then a Gunn-effect domain. The growth rate is given by the negative dielectric relaxation time $\tau_n = 1/\omega_{dn}$. A domain at its early stages will therefore grow fastest if $n_0$ and the negative differential mobility are largest.

A domain usually grows to a large size. $V_D$ takes more and more of the bias voltage $V_B$ applied to the ohmic electrodes of the device, until the field outside the domain has dropped to a value which is well below the threshold field for the transferred electron effect. Then the domain has reached a steady-state condition. The shape of the domain field distribution is now approximately triangular, as established by careful probe measurements and by numerical computation. The front part is given by depletion of carriers, whereas the back part is an accumulation layer of heavy satellite-valley electrons. One can understand how a small initial field perturbation produces a bunch of heavy electrons which cannot keep up with the light electrons. A depletion layer is created which generates a high field in accordance with Poisson's equation, with the result that all electrons generated there will immediately become satellite valley electrons. It can therefore be understood that this layer is almost completely depleted. Assuming, then, complete depletion, the excess field of a mature domain increases steadily from the domain front at $x = 0$ by

$$E_d = \frac{en_0x}{\varepsilon}$$

until the accumulation layer is reached at $x = d_a$. The accumulation layer has to bring $E_d$ down to zero again. As the accumulation density $n_a$ is determined by the diffusion constant, the thickness $d_a$ of this layer can be found again by taking the maximum value of $E_d$, i.e.

$$d_a = \left(\frac{en_0d_d}{\varepsilon}\right)\frac{\varepsilon}{en_a}$$

The value of $d_a$ is approximately constant and is determined by the diffusion constant $D$. For large domains, it can be about one-tenth of $d_d$, so that one can write approximately

$$V_D = \tfrac{1}{2}en_0d_d^2/\varepsilon$$

The field outside the domain, $E_R$, will then form with the low-field resistance, $R_0$, of the device, and the interelectrode distance $l$, the diode current $I_D$ during transit of a mature domain, i.e.

$$I_D = \frac{E_Rl}{R_0}$$

When the domain reaches the anode electrode, it collapses there and the diode current increases again. When the threshold current $I_t$ for domain nucleation, and thus the threshold field at the nucleating centre, has been reached, a new domain is formed again. When a sufficiently large bias voltage is applied, one can obtain current oscillations whose amplitude is approximately $\tfrac{1}{2}(I_t - I_D)$, and whose frequency is given by the inverse of the domain transit time. Of course, the interelectrode distance must be large enough for domains to grow to a reasonable size. This yields the condition that $\omega_d l/v_D \gg 1$, which has often been expressed as

$$nl > 10^{11}/\text{cm}^2 \tag{2.9}$$

($v_D$ is the domain velocity which is usually around $10^{-7}$ cm/s).

## 2.2 THE MATURE DOMAIN

In order to treat a fully grown domain mathematically, we transform equation (2.5) so that the origin of the co-ordinates move with the domain.

We therefore introduce the new co-ordinates

$$z' = z - \int_0^t u(t)\, dt$$

and

$$t' = t$$

so that

$$\frac{\partial}{\partial t} = \frac{\partial}{\partial t'} - u\frac{\partial}{\partial z'}$$

and

$$\frac{\partial}{\partial z} = \frac{\partial}{\partial z'}$$

With the new co-ordinates equation (2.5) becomes

$$\frac{I}{A} = env(E) - eD\frac{\partial n}{\partial z} - u\varepsilon\frac{\partial E}{\partial z} + \varepsilon\frac{\partial E}{\partial t} \tag{2.10}$$

where the primes of the new co-ordinates have been dropped again. We now consider the variable to be $E$ and not $z$, which can be expressed with the help of Poisson's equation (equation (2.6) where $z$ is the new co-ordinate $z'$ now), i.e.

$$\frac{\partial}{\partial z} = \frac{\partial}{\partial E} \cdot \frac{\partial E}{\partial z} = \left[\frac{e(n - n_0)}{\varepsilon}\right]\frac{\partial}{\partial E} \tag{2.11}$$

Additionally we only consider the steady state so that $\partial/\partial t = 0$ in equation (2.10), which becomes, using (2.11),

$$\frac{I/A - en_0 u}{en} = [v(E) - u] - \frac{De}{\varepsilon n}\frac{dn}{dE}(n - n_0)$$

This is then integrated from the field outside the domain, $E_\mathrm{R}$, to the peak field $E_\mathrm{M}$:

$$\frac{en_0 D}{\varepsilon}\int_{n(E_\mathrm{R})}^{n(E_\mathrm{M})}\left(1 - \frac{n_0}{n}\right)\frac{dn}{n_0} = \int_{E_\mathrm{R}}^{E_\mathrm{M}}[v(E) - u]\, dE + \int_{E_\mathrm{R}}^{E_\mathrm{M}}\frac{I/A - en_0 u}{en}\, dE \tag{2.12}$$

$E_\mathrm{M}$ (where $\partial E/\partial z = 0$) occurs at the point where $n = n_0$ as the depletion changes over to accumulation, so that $n(E_\mathrm{R}) = n(E_\mathrm{M})$ and the left-hand side of equation (2.12) disappears. The first integral on the right-hand side of (2.12) is only a function of $E$ and is therefore independent of the path taken for the integration, that is, for the integration from $E_\mathrm{R}$ to $E_\mathrm{M}$ via the accumulation or via the depletion layer. The last integral has a variable $1/n$, which depends on the path of integration taken, as $n$ is small in the depletion layer and large in the accumulation part of the domain. The integral equation (2.12) can therefore only be satisfied if the last integral term is always zero, which means that $I/A - en_0 u$ must be zero. This indicates that the velocity of the electrons outside the domain is equal to the domain velocity, as $I/A = en_0 v(E_R) = en_0 u$ with $v(E_R) = u$. Finally, one has the integral equation

$$\int_{E_R}^{E_M} [v(E) - u]\, dE = 0 \tag{2.13}$$

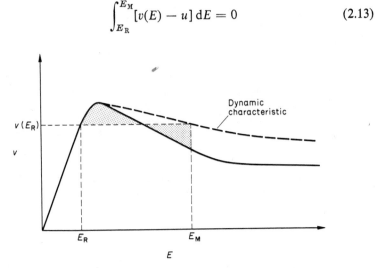

**Figure 2.3**   Velocity-field characteristics showing equal areas (shaded parts) and dynamic characteristic.

This means that the domain velocity $u$ is determined by the condition for equal shaded areas in Figure 2.3. This is because equation (2.13) says, with

$$\int_{E_R}^{E_M} v(E)\, dE = u(E_M - E_R)$$

that $u$ and $E_R$ are determined for each value of $E_M$ by the equal-areas rule, which produces a dynamic characteristic as introduced in Figure 2.3 by a dashed line.

Note that the analysis does not need to make the approximation of zero diffusion current as long as $D$ is a field-independent constant. If a field-dependent $D$ is taken into account, a more complex rule involving unequal areas is found.[5] In this case the domain moves at a slightly different velocity from that of the external electrons.

As we have shown that the last term of equation (2.12) vanishes because the numerator is a constant which is zero, we can integrate this equation from $E_R$ to some general field value $E$ between $E_R$ and $E_M$. This produces the following relation:

$$\left(\frac{n}{n_0} - \log_e \frac{n}{n_0} - 1\right) = \frac{\varepsilon}{Den_0} \int_{E_R}^{E} [v(E) - v(E_R)]\, dE \tag{2.14}$$

where we have set $n(E_R) = n_0$. If $v(E)$ is known, the integral on the right-hand side can be evaluated. The function $F$ as given by the left-hand side of equation (2.14) can easily be evaluated too and one finds that there are always two values of $n/n_0$ producing the same value of $F$. In fact, there is always one value of $n$ larger than $n_0$ and one value less than $n_0$ for every value of $F$, corresponding to respectively the accumulation and depletion layers of the domain. The numerical evaluation of $F$ shows that for small $D$ the depletion is almost complete, whereas accumulation is very high. Then the domain approaches a triangular shape and the excess domain voltage $V_D$ can be obtained approximately from Poisson's equation as shown in the previous section, by assuming that the accumulation layer is much smaller than the depletion layer which is considered to be entirely depleted. The resulting expression was found by double integration of Poisson's equation, i.e.

$$V_D = \frac{\frac{1}{2}en_0 d_d^2}{\varepsilon} \tag{2.15}$$

By a single differentiation, the maximum excess field in the domain is found to be

$$E_M - E_R = \frac{en_0 d_d}{\varepsilon} \tag{2.16}$$

which yields, when introduced into equation (2.15),

$$V_D = \frac{\varepsilon(E_M - E_R)^2}{2en_0} \tag{2.17}$$

The maximum excess field can be obtained as a function of rest field $E_R$ with the help of the equal-areas rule.

An often-employed approximation for the $v(E)$ characteristic is a three-line concept as shown by Figure 2.4. The conditions for the equal-areas rule can be written down easily then, i.e. for $E_M > E_s$,

$$(E_t - E_R)^2\mu_0 + (E_p - E_t)^2\mu_n = (E_s - E_p)^2\mu_n + (E_s - E_p)(E_M - E_s)\mu_n$$

and for $E_M < E_s$,

$$(E_t - E_R)^2\mu_0 + (E_p - E_t)^2\mu_n = (E_M - E_p)^2\mu_n$$

where all terms are defined in Figure 2.4.

Using the relations

$$(E_p - E_t)\mu_n = (E_t - E_R)\mu_0 \tag{2.18}$$

and

$$E_s = E_t + v_t(1 - k)/\mu_n \tag{2.19}$$

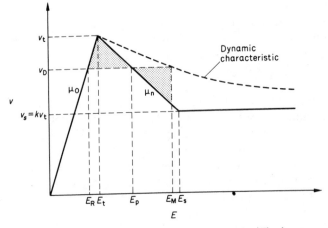

**Figure 2.4** Three-line approximation of $v(E)$ characteristics and dynamic characteristic

to eliminate $E_p$ and $E_s$, one finds the following expressions: for $E_M < E_s$,

$$E_M - E_R = \left[\frac{\mu_0}{\mu_n}\left(\frac{\mu_n}{\mu_0} + 1\right)^{\frac{1}{2}} + \frac{\mu_0}{\mu_n} + 1\right](E_t - E_R) \qquad (2.20a)$$

For $E_M > E_s$,

$$E_M - E_R = \frac{1}{2}\left(\frac{\mu_0}{\mu_n} + 1\right)\frac{(1 - k)^2 E_t^2 - \frac{1}{2}(E_R - kE_t)}{(E_R - kE_t)} \qquad (2.20b)$$

Here the parameter $k$ is equal to $v_s/v_t$ whose inverse is often called the peak-to-valley ratio. This is usually used to donote the device-current ratio for current oscillations due to mature domains, when the ratio $1/k$ is of course equally applicable.

Inserting equation (2.20) into equation (2.17) then gives the relation between excess domàin voltage $V_D$ and rest field $E_R$ for a given semi-conducting material, which is described by the parameters $\mu_0$, $\mu_n$, $k$, and $E_t$. Taking $\mu_n/\mu_0 = 0.3$, $k = 0.4$, and $E_t = 3000$ V/cm, which are typical values for GaAs, the dependence of Figure 2.5 is obtained. Here, only those $E_R$ values are of interest which are smaller than $E_t$, as can be seen from Figure 2.4. For $E_R$ near $E_t$, equation (2.20a) is employed until $E_M = E_s$, which occurs according to equations (2.19) and (2.20a) at

$$E_R = E_t\left\{+\frac{\mu_0}{\mu_n}\left[\frac{\mu_n}{\mu_0} + 1\right]^{\frac{1}{2}} + \frac{\mu_0}{\mu_n} - (1 - k)\frac{\mu_0}{\mu_n}\right\} \times \left[\frac{\mu}{\mu_n}\left[\frac{\mu_n}{\mu_0} + 1\right]^{\frac{1}{2}} + \frac{\mu_0}{\mu_n}\right]^{-1}$$

$$= E_t\, 0.715 \qquad (2.21)$$

Then equation (2.20b) is relevant. The minimum value of $E_R$ occurs, of course, at $E_R = kE_t$, when $V_D$ is infinitely large.

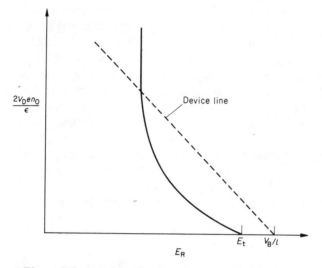

**Figure 2.5** Domain characteristics excess voltage $V_D$ versus electric field outside domain $E_R$. The dashed line is the device line which gives the values of $V_D$ and $E_R$ by the cross point.

As can be seen from equation (2.17), the value of $V_D$ is inversely proportional to the carrier density (which is equal to $n_0$).

The details of Figure 2.5 have usually been presented in this way for Gunn-effect domains, although it gives the same details as the commonly known current-voltage characteristics, since $E_R$ is proportional to the instantaneous current $I$ ($= An e \mu_0 E_R$) during domain transit whereas $V_D$ is the instantaneous device voltage minus $E_R l$.

Figure 2.5 can now be employed to give the value of $V_D$ for a given length $l$. For this purpose the device line is entered into this graph as indicated by a dashed line, that is, the line starts on the $E_R$-axis at a value which is the field $V_B/l$; it then has a slope which is given by the value of $l$ in accordance with the voltage/field axes. A device line shows also how a Gunn-effect device with a bias voltage slightly below $E_t l$ can be made to carry a stable domain by applying a short voltage pulse which increases the field temporarily above $E_t$. One can see from Figure 2.5 the minimum bias voltage for a given value of $l$, i.e. the smaller $V_B$ is, the larger must be $l$. It is possible, in fact to derive again a minimum $nl$ product for sub-threshold biased diodes to be able to carry domains, by using equations (2.20) and (2.17). For example, for $E_M < E_s$, one obtains

$$(nl)_{st} = \frac{\varepsilon}{e}\left(\frac{\mu_0}{\mu_n}\left[\frac{\mu_n}{\mu_0} + 1\right]^{\frac{1}{2}} + \frac{\mu_0}{\mu_n} + 1\right)(E_t - E_R) \qquad (2.22)$$

where $st$ denotes the limiting product for subthreshold biasing.

It should be pointed out here that these considerations are only of qualitative value, because the approximations made in setting up equation (2.17) assume large mature domains, whose fully depleted depletion layers are very much larger than the accumulation layer. This means that near $E_R = E_t$, the quantitative details are not necessarily justifiable.

When $D$ is not small, as employed for the above approximate analysis, the values of $n$ in the accumulation and depletion layers of the domain have to be calculated by using equation (2.14). Again using the three-line approach for the $v(E)$ characteristics, one can approach this problem by limiting the voltage either to high values or to low ones.

In the latter case the change in carrier density is assumed to be small. The function $F$ given by the left-hand side of equation (2.14) is then expanded by a Taylor series, retaining only the first term, i.e.

$$F \simeq \frac{1}{2}\left(\frac{n - n_0}{n_0}\right)^2 \tag{2.23}$$

The integral $I_n = \int_{E_R}^{E} [v(E) - E_R\mu_0]\, dE$ of equation (2.14) can be evaluated again by writing down the expression for the surface of the triangles as given by the three-line approximation for $v(E)$ (Figure 2.4). This is a similar approach to that used by the equal-areas rule, except that the areas involved are, of course, not necessarily equal if the integration is from $E_R$ to some general field value $E$ of the domain. $I_n$ then gives for $E_R < E < E_t$,

$$I_n = \tfrac{1}{2}\mu_0(E - E_R)^2 \tag{2.24}$$

and for $E_t < E < E_s$,

$$I_n = \frac{1}{2}\mu_0(E_t - E_R)^2 + \frac{1}{2}\frac{\mu_0^2}{\mu_n}(E_t - E_R)^2 \pm \frac{1}{2}\mu_n(E - E_p)^2 \tag{2.25}$$

with $E_p = E_t + (E_t - E_R)(\mu_0/\mu_n)$ (see equation (2.18)). The minus sign in front of the last term in equation (2.25) is for $E_R < E < E_p$, whereas the plus sign counts for $E_p < E < E_s$. As we are treating here the approximation of small domain voltages, we do not consider the case of $E > E_s$.

Poisson's equation

$$\frac{dE}{dz} = \frac{(n - n_0)e}{\varepsilon} \tag{2.26}$$

is inserted into equation (2.23), giving for equation (2.14), with $E_R < E < E_t$,

$$\pm\left[\frac{D\varepsilon}{en_0\mu_0}\right]^{\frac{1}{2}}\frac{dE}{dz} = E - E_0 \tag{2.27}$$

and with $E_t < E < E_s$,

$$\pm\left[\frac{D\varepsilon}{en_0\mu_0}\right]^{\frac{1}{2}}\frac{dE}{dz} = \left[(E_T - E_R)^2 + \frac{\mu_0}{\mu_n}(E_t - E_R)^2 \mp \frac{\mu_n}{\mu_0}(E - E_p)^2\right]^{\frac{1}{2}} \tag{2.28}$$

Equation (2.27) can be solved easily by using the ansatz of an exponential function, namely

$$E - E_{\mathrm{R}} = (E_t - E_{\mathrm{R}}) \exp \left( \pm \frac{(z - z_{1,2})}{L_{\mathrm{D}}} \right) \tag{2.29}$$

where

$$L_{\mathrm{D}} = \left[ \frac{D\varepsilon}{e n_0 \mu_0} \right]^{\frac{1}{2}} \tag{2.30}$$

known as the Debye length. Here $z_1$ and $z_2$ are the two points at the front and at the back of the domain where the threshold field is reached. This solution shows how the field rises exponentially for $E_{\mathrm{R}}$ at both sides of the domain, until at $z_1$ and $z_2$ the solution of equation (2.28) is relevant.

Equation (2.28) can easily be solved by employing the following ansatz:

$$E - E_{\mathrm{R}} = (E_t - E_{\mathrm{R}}) \left[ A_\mu + B_\mu \cos \left( \frac{z}{L_{\mathrm{S}}} \right) \right] \tag{2.31a}$$

Inserting this expression into equation (2.28) yields

$$\pm \frac{L_{\mathrm{D}}}{L_{\mathrm{S}}} (E_t - E_{\mathrm{R}}) B_\mu (-\sin (z/L_{\mathrm{S}}))$$

$$= \left\{ (E_t - E_{\mathrm{R}})^2 \left( 1 + \frac{\mu_0}{\mu_n} \right) \right.$$

$$\left. \mp \frac{\mu_n}{\mu_0} \left[ (E_t - E_{\mathrm{R}}) \left( A_\mu + B_\mu \cos \frac{z}{L_{\mathrm{S}}} \right) + E_{\mathrm{R}} - E_t - \frac{\mu_0}{\mu_n} (E_t - E_{\mathrm{R}}) \right]^2 \right\}^{\frac{1}{2}}$$

or

$$\left( \frac{L_{\mathrm{D}}}{L_{\mathrm{S}}} B_\mu \right)^2 \sin^2 \frac{z}{L_{\mathrm{S}}} = 1 + \frac{\mu_0}{\mu_n} \mp \frac{\mu_n}{\mu_0} \left[ A_\mu + B_\mu \cos \frac{z}{L_{\mathrm{S}}} - 1 - \frac{\mu_0}{\mu_n} \right]^2$$

This equation can be satisfied if

$$A_\mu = 1 + \frac{\mu_0}{\mu_n}$$

The remaining equation can be reduced to

$$\left( \frac{L_{\mathrm{D}}}{L_{\mathrm{S}}} B_\mu \right)^2 \sin^2 \frac{z}{L_{\mathrm{S}}} \pm \frac{\mu_n}{\mu_0} B_\mu^2 \cos^2 \frac{z}{L_{\mathrm{S}}} = 1 + \frac{\mu_0}{\mu_n}$$

and finally becomes a well-known trigonometric relation for

$$L_{\mathrm{S}} = L_{\mathrm{D}} \left[ \frac{\mu_0}{\mu_n} \right]^{\frac{1}{2}} \quad \text{and} \quad B_\mu = \left[ \frac{\mu_0}{\mu_n} + \left( \frac{\mu_0}{\mu_n} \right)^2 \right]^{\frac{1}{2}}$$

if the positive sign is used, i.e. if $E_{\mathrm{R}}$ is near $E_t$, so that $E$ only lies between $E_t$ and $E_{\mathrm{p}}$ for a very small distance. Otherwise the solution has to take the following hyperbolic form:

$$E - E_R = (E_t - E_R)\left[A_\mu + B_\mu \sinh \frac{z - z'_{a,b}}{L_S}\right] \qquad (2.31b)$$

where constants $A_\mu$, $B_\mu$, and $L_S$ have the same values as above, as can be found by inserting this solution into equation (2.28).

Equation (2.31a) is now valid up to the two values of $z$ where $E = E_p$. At those two points, equation (2.31b) takes over. The values of field $E$ must be equal there. Therefore the two values $z'_a$ and $z'_b$ have to be such that $E = E_p$ in equation (2.31b). Finally, at $E = E_t$, equation (2.29) has to take over. $z_1$ and $z_2$ are found by placing $E = E_t$ in equation (2.31b), so that the field values are again equal for both equations (2.29) and (2.31b) at these two points.

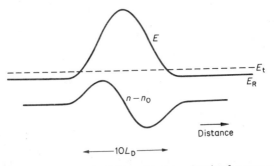

**Figure 2.6** Domain field and carrier density for case of low-voltage limit.

The carrier density $n$ in the domain can be determined by using equation (2.26). Both the domain field and electron density are shown in Figure 2.6 for a low-voltage domain with non-negligible diffusion. Of course, the $n$-dependence must also be continuous, although one can expect some difficulties here, in particular at the points of change from equation (2.29) to (2.31b), as the carrier density shows a discontinuity there. Fortunately the importance of both equations (2.29) and (2.31b) is small as $E_R$ is very close to $E_t$ (and therefore $E_p$ is close to $E_t$) for small values of domain voltages. It is thus possible to employ primarily the solution of (2.31a) for the carrier density distribution, and to neglect the domain edge details as given by (2.29) and (2.31b).

So far we have not yet fixed the amplitude of $E_R$. This is determined by finding $V_D$ from the equal-areas rule as shown above because the integration of equation (2.14) is then performed from $E_R$ to $E_M$, where for a constant $D$, $n_0 = n$ so that the left-hand side of equation (2.14) becomes zero again. One obtains a characteristic of $V_D = f(E_R)$ again which fixes $E_R$ for a given device line due to $l$ and bias voltage (see Figure 2.5). For a given problem, one would of course first determine $E_R$ and subsequently $E(z)$ and $n(z)$.

Figure 2.6 and equations (2.29) and (2.31) demonstrate the interesting result that a small domain is symmetrical. The peaks of the accumulation and depletion layer are separated by $L_S = L_D(\mu_0/\mu_n)^{\frac{1}{2}}$. The Debye length is about 0.2 μm in 1 Ω cm GaAs material. This illustrates the actual size of a small domain. The best ratio of $\mu_0/\mu_n$ is around 3 so that $L_S$ is at least $1.7L_D$. However, inferior material or elevated lattice temperatures $T_0$ (see Figure 2.1, page 6) decrease $\mu_n$ so that $L_S$ can become very large until the domain becomes unstable and disappears.

The domain voltage can also be obtained from equations (2.29) and (2.31) by integrating the excess field, i.e.

$$V_D = \int (E - E_R)\, dz$$

Using equation (2.31a) only, the result can be obtained easily, giving:

$$V_D \cong 2L_D \left(\frac{\mu_0}{\mu_n}\right)^{\frac{1}{2}} \left[ (E_t - E_R)\left(1 + \frac{\mu_0}{\mu_n}\right) + \left(\frac{\mu_0}{\mu_n} + \frac{\mu_0^2}{\mu_n^2}\right)^{\frac{1}{2}} \right] \tag{2.32}$$

The maximum field $E_M$ occurs in the centre where $\cos z/L_S = 1$ in equation (2.31a),

$$E_M - E_R = (E_t - E_R)\left\{ 1 + \frac{\mu_0}{\mu_n} + \left[ \frac{\mu_0}{\mu_n} + \left(\frac{\mu_0}{\mu_n}\right)^2 \right]^{\frac{1}{2}} \right\} \tag{2.33}$$

As the charge stored in the accumulation layer is

$$Q = A\varepsilon(E_M - E_R)$$

the domain voltage is

$$V_D = 2L_D \left(\frac{\mu_0}{\mu_n}\right)^{\frac{1}{2}} \left\{ \frac{Q}{A\varepsilon[1 + (1 + \mu_n/\mu_0)^{-\frac{1}{2}}]} + \left(\frac{\mu_0}{\mu_n} + \frac{\mu_0^2}{\mu_n^2}\right)^{\frac{1}{2}} \right\} \tag{2.34}$$

The domain voltage is now linearly related to the charge stored, which means the domain capacitance is approximately constant and independent of $V_D$, in contrast to the fully grown domain, whose domain capacitance is approximately

$$C_D = \frac{\varepsilon A}{d_d} = \frac{\varepsilon A}{(2V_D\varepsilon/en_0)^{\frac{1}{2}}} = A\left(\frac{\varepsilon en_0}{2V_D}\right)^{\frac{1}{2}} \tag{2.35}$$

Finally it is shown how the steady-state domain can be treated analytically by using a high-voltage approximation. Most of the domain field will then be above $E_s$, and $u \cong v_s$, so that one obtains from equation (2.10), using the same arguments regarding $I/A - en_0u = 0$ as employed for equation (2.13), the following expression:

$$0 = v_s - v(E_R) - \frac{D}{n} \frac{dn}{dz} \tag{2.36}$$

Poisson's equation (equation (2.6), page 8) gives after integration:

$$E - E_M = \frac{e}{\varepsilon} \int_0^z (n - n_0) \, dz$$

$$= \frac{e}{\varepsilon} \int_0^z n \, dz - \frac{e n_0}{\varepsilon} z \tag{2.37}$$

Equation (2.36) can be rewritten

$$n \, dz = \frac{-D}{v(E_R) - v_s} \, dn$$

$$\frac{1}{D} \int_0^z (v(E_R) - v_s) \, dz = - \int_0^n \frac{1}{n} \, dn \tag{2.38}$$

or, after integrating both sides,

$$\exp - \frac{(v(E_R) - v_s)z}{D} = n \tag{2.39}$$

Inserting equation (2.38) into (2.37), and changing the boundary conditions of the integral correspondingly by using (2.39), gives

$$E(z) - E_M = \frac{e D n_0}{\varepsilon (v(E_R) - v_s)} \left[ 1 - \exp - \frac{v(E_R) - v_s}{D} z \right] - \frac{e n_0 z}{\varepsilon} \tag{2.40}$$

Here $z = 0$ at $E = E_M$. The solution extends from $z = -z_a$ on the accumulation side to $z = +z_d$ on the depletion side. These points lie where $E = E_s$. $E_p$ is, of course, still marginally below $E_s$. $E_p$ occurs where the diffusion effects on the electron velocity are zero, i.e. $\partial n / \partial z = 0$. This means that these are the points where the carrier density has reached a maximum or minimum. Figure 2.7 shows the field distribution and carrier density, as obtained from equation (2.40) by differentiation, and from equation (2.6), whereas Figure 2.8 summarizes again the field definitions used in this analysis.

From the field profile, the voltage $V_D$ can be found by integration again. This is not attempted here, however, as the analysis is very lengthy and tedious. The main conclusion is that $V_D$ is proportional to $L_D$ rather than the transit length $v_t / \omega_d$, as can be shown from equation (2.17). At high resistivities the diffusion-free model should be applicable; however even up to 10 $\Omega$ cm material, diffusion effects are noticeable. The lowest limit of resistivity must be somewhere below 0.1 $\Omega$ cm, and is given by

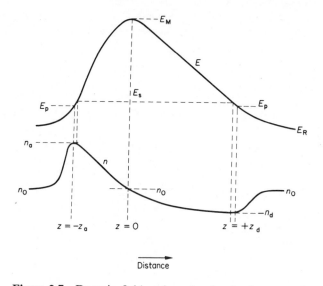

**Figure 2.7** Domain field and carrier density for case of high-field limit.

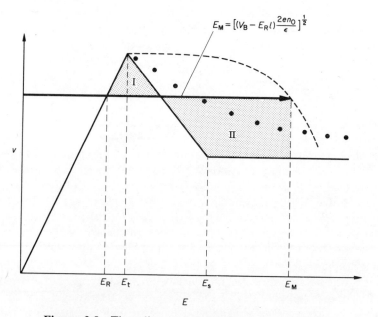

$$E_M = \left[(V_B - E_R l)\frac{2en_0}{\epsilon}\right]^{\frac{1}{2}}$$

**Figure 2.8** Three-line approximation of $v(E)$ with domain amplitude function for fully grown domains (dotted line) and for the transient case (broken line) $\mu_0$ = low-field mobility, $\mu_n$ = negative mobility.

the fact that impurity scattering becomes very important and begins to inhibit the electron transfer mechanism.

## 2.3 THE TIME DEPENDENCE OF DOMAINS

Growth and decay of domains are important phenomena, particularly in connection with pulse applications as they determine the pulse times achievable with the transferred-electron effect. The transient analysis employs equation (2.10) in a similar manner as the steady state treatment described above, except that $\partial/\partial t \neq 0$, so that

$$0 = \int_{E_R}^{E_M} (v(E) - u)\, dE + \int_{E_R}^{E_M} \frac{1}{en} \left( \frac{I}{A} - en_0 u - \varepsilon \frac{\partial E}{\partial t} \right) dE \quad (2.41)$$

Here the fact is used again that $n = n_0$ at $E = E_M$ even in the changing domain. Previously it was argued that $I/A - en_0 u$ must be zero, as otherwise the right-hand integral would be dependent on the integration path (see arguments in connection with equation (2.12)). This meant that the external-electron velocity equals that of the domain. Now for the transient case, this is no longer correct as a growing domain moves more slowly than the external electrons, thus increasing both the width of the depletion layer and collecting more electrons in the accumulation layer. Similarly the decaying domain moves faster than the external charge carriers. By rewriting the right-hand integral of equation (2.41) one obtains

$$I_n = \int_{E_R}^{E_M} \left[ en_0(v(E_R) - u) - \frac{\partial Q(z)}{\partial t} \right] \frac{1}{en}\, dE \quad (2.42)$$

with $Q = \varepsilon(E - E_R)$, the excess charge per unit area, stored between point $z$ with field $E$ and $z(E_R)$. The reader is reminded that $z$ was previously defined as a co-ordinate whose origin moves with the domain, as introduced at the beginning of section 2.2, page 11, for the derivation of equation (2.10). With such a moving frame of reference, it is of course no more possible to say, as was done in connection with equation (2.12), that the integral of equation (2.42) always is zero, because $\partial Q/\partial t$ is not a constant and is indeed different for the front and back part of the domain so that the integral can remain independent of integration path. However, one can approach this problem in the following way. We take a moving point $z_q$ where no charge is crossing during domain charging or discharging. This $z_q$ moves towards $z(E_M)$ during charging and away from $z(E_M)$ during discharging. One can in fact define two such points, one in front of $z(E_M)$ and one behind. For the $z_q$ behind $z(E_M)$, for example, one can say the following, (using the domain representation of Figure 2.7, page 21): charge is moving into the left at a rate $en_0(v_R - u)$ for domain growth and this increases $Q(z)$, whereas to the right of $z_q$ up to $z(E_M)$ no charge is entered but the distance $z_q - z(E_M)$ is contracted for domain growth, thus increasing $\partial Q/\partial t$ there all the same. The integrand is then zero at $z_q$ and changes sign there. The integral $I_n$ (equation 2.42) can then be separated into the following two parts:

$$I_n = \int_{E_R}^{E_{(z_q)}} + \int_{E_{(z_q)}}^{E_M}$$

If the correct values of $z_q$ are taken, the two parts of $I_n$ are equal in magnitude and opposite in sign so that $I_n = 0$. This then means that the first term of the right-hand side of equation (2.41) is zero again and defines another equal-areas rule.

To summarize the analytical procedure one has first to find $z_q$ and its velocity $u$ from the condition that the above two parts of $I_n$ are zero. Then, using the new equal-areas rule, where $u$ is no more equal to $v(E_R)$, one determines the device behaviour.

As diffusion effects decrease or the domain voltage $V_D$ increases, the points $z_q$ move closer and closer to $z(E_M)$ so that it is often reasonable to take as $u$ the velocity of the point $z(E_M)$. Then the total charge stored in the accumulation layer, $Q$, is given approximately by the relation

$$\frac{dQ}{dt} = en_0(v(E_R) - u) \tag{2.43}$$

where $Q = \varepsilon(E - E_R)$ again. Using the three-line characteristic of Figure 2.4, one finds for $E_s \geqslant E_R \geqslant E_t$ and $E_s \geqslant E_M \geqslant E_t$,

$$u - v(E_R) = -\tfrac{1}{2}\mu_n(E_M - E_R) \tag{2.44a}$$

for $E_s > E_M > E_t$ and $E_R < E_t$,

$$u - v(E_R) = (\mu_0 + \mu_n)(E_t - E_R) - \tfrac{1}{2}\mu_2(E_M - E_R)$$
$$- \frac{\tfrac{1}{2}(\mu_0 + \mu_n)(E_R - E_t)^2}{(E_M - E_R)} \tag{2.44b}$$

and for $E_M > E_s$ and $E_R < E_t$,

$$u = v_s + \left[ \frac{\{(\mu_0 + \mu_n)E_t v_t(1 - k)^2 - \mu_0\mu_n(E_R - kE_t)^2\}}{2\mu_n(E_M - E_R)} \right] \tag{2.44c}$$

If a local resistivity peak or a momentary bias spike has nucleated a domain, the instability will grow further by using equations (2.44a) with (2.43). One then obtains

$$\frac{dQ}{dt} = \tfrac{1}{2}\omega_{dn}Q \tag{2.45}$$

where $\omega_{dn} = en_0\mu_n/\varepsilon$. The charge in the accumulation layer increases as $\exp(\tfrac{1}{2}\omega_{dn}t)$. The domain voltage grows in accordance with the following relation:

$$V_D = \int_{-\infty}^{+\infty} (E - E_R)\, dz = \frac{1}{\varepsilon} \int_{-\infty}^{+\infty} Q\, dz$$

$$= \frac{1}{\varepsilon} \int_{-\infty}^{+\infty} Q_0\, dz \exp \omega_{dn} t$$

$$= V_{DO} \exp \omega_{dn} t \tag{2.45}$$

which is in agreement with the solution of equation (2.8). $V_{DO}$ is the initial voltage perturbation which is required to nucleate a domain.

Once $E_M = E_s$, the growth slows down. For the approximation $E_R \cong E_t$, and for $E_M > E_s$, the following new relation can be derived:

$$\frac{1}{\omega_{do}} \frac{dQ}{dt} = (1 - k)Q_t - \tfrac{1}{2}(1 - k)^2 \left(\frac{\mu_0}{\mu_n}\right) \frac{Q_t^2}{Q} \tag{2.47}$$

with $\omega_{do} = e n_0 \mu_0 / \varepsilon$ and $Q_t = \varepsilon E_t$. The solution of equation (2.47) is given by the following equation:

$$Q - Q_1 + gQ_t \log_e \left[(Q - gQ_t)/(Q_1 - gQ_t)\right]$$
$$= (1 - k)\omega_{do} Q_t (t - t_1) \tag{2.48}$$

with $g = \tfrac{1}{2}(1 - k)\mu_0/\mu_n$, and where $Q_1$ is the charge at time $t_1$ when the field first reaches $E_s$. Equation (2.48) shows that the domain charge grows then less than linearly owing to the occurrence of $Q$ in the third term on the left-hand side. Equation (2.48) represents in fact a very fast growth rate, particularly as the high external field (with the above assumption of $E_R \cong E_t$) transfers a maximum amount of charge into the accumulation layer. It is stressed that the solution of equation (2.48) is given for maximum growth rates with $E_R \cong E_t$ whereas usually $E_R < E_t$ and decreases with increasing domain charge. As soon as the bias voltage cannot maintain the condition for a high value of $E_R$ any more, domain growth will be slowed down considerably until the steady-state condition is reached asymptotically.

It can be shown that for a constant bias voltage $V_B$ applied to the device, the maximum charge $Q_m$ is approached approximately as

$$Q = Q_m \tanh \left[\frac{Q_t}{Q_m}\right] (1 - k)\omega_d t \tag{2.49}$$

It is possible to derive from equation (2.7) an expression which permits one to estimate domain growth and decay without having to depend on a small-signal approximation as required for the derivation of equation (2.8). One has to assume that $D$ is small. The accumulation layer is considerably shorter than the leading edge where additionally full depletion is assumed; using equation (2.6), one finds

$$\int_{z_2}^{z_1} \frac{n_0 e}{\varepsilon} \{v(E_R) - v(E)\} \, dz \simeq \int_{E_R}^{E_M} \{v(E_R) - v(E)\} \, dE$$

Substituting this into equation (2.7) and remembering that the last term of (2.7) is zero, one obtains

$$\frac{dV_D}{dt} \simeq \int_{E_R}^{E_M} \{v(E_R) - v(E)\} \, dE \tag{2.50}$$

For the steady state, $dV_D/dt = 0$ and equation (2.50) is the same as equation (2.13), which defines the equal-areas rule as $v(E_R) = u$. Equation (2.50) can be employed to define an unequal-areas rule. If the lower right-hand shaded area of Figure 2.3, page 12 is smaller or larger than the upper shaded area, the domain decays or grows respectively. The difference gives the rate of change.

At this stage one can show also that a Gunn-effect device biased with a constant bias current cannot have a stable domain. When the bias current is increased from zero to the threshold value, where $v = v_t$ (see Figure 2.4, page 14), a domain is nucleated then and continues to grow at an ever-increasing rate; this is because $v(E_R)$ cannot decrease owing to the constant-current supply, and so the inequality of the shaded areas as shown by Figure 2.4 increases. This consideration also demonstrates that the series resistance of a Gunn-effect element must not be too high in order to prevent the domain field $E_M$ growing dangerously large. A danger level is in fact reached when $E_M$ approaches the value of field ionization of electron-hole pairs.

The unequal-areas rule defined by equation (2.50) can be employed to derive a relatively simple analytical expression for domain growth which does not rely on the approximation $E_R \simeq E_t$ as required for equations (2.47) and (2.48). This gives a useful expression which shows clearly the effect of a load resistance on domain growth rates. Rewriting equation (2.50), one finds the growth time $T_g$ for a domain to achieve a voltage $V_{Dm}$:

$$\int_{V_D=0}^{V_{Dm}} \frac{1}{\int_{E_R}^{E_M} [v(E_R) - v(E)] \, dE} \, dV_D = T_g \tag{2.51}$$

Except for very small domains and for an appreciable diffusion constant, one has equation (2.17), i.e.

$$\frac{(E_M - E_R)^2 \varepsilon}{2 e n_0} = V_D = V_B - E_R l \tag{2.52}$$

Using the three-line approximation again for $v$ vs. $E$, we can enter the lines for the domain amplitude $E_M$ of fully grown domains using the equal-areas rule (dotted line in Figure 2.8, page 21) and for the transient case using equation (2.52) (broken line in Figure 2.8). The integral of the denominator of equation (2.51) gives the difference between the areas I and II as shown by shading in Figure 2.8, and can therefore be expressed as follows for $E_t < E_M < E_s$:

$$F = \frac{1}{2}(E_t - E_R)^2 \left(\mu_0 + \frac{\mu_0^2}{\mu_n}\right) - \frac{1}{2}\left[E_M - E_R - (E_t - E_R)\left(1 + \frac{\mu_0}{\mu_n}\right)\right]^2 \mu_n$$

(2.53a)

and for $E_M > E_s$,

$$F = \frac{1}{2}(E_t - E_R)^2 \left(\mu_0 + \frac{\mu_0^2}{\mu_n}\right) - \frac{1}{2}\left[-E_t \frac{k\mu_0}{\mu_n} + E_R \frac{\mu_0}{\mu_n}\right]^2 \mu_n$$
$$- (E_M - E_s)\left(E_t \frac{k\mu_0}{\mu_n} + E_R \frac{\mu_0}{\mu_n}\right)\mu_n \quad (2.53b)$$

Re-arranging and using equation (2.52) for $E_M$, this gives the following type of series:

$$F = \sum_{n=0}^{4} a_n E_R^{\frac{1}{2}n}$$

(2.54)

Inserting equation (2.54) into (2.51) shows that an integral of the following form has to be solved:

$$-l \int_0^{V_{Dm}} \frac{dE_R}{\sum_{n=0}^{4} a_n E_R^{\frac{1}{2}n}}$$

(2.55)

where we used from equation (2.52), $dV_D/dE_R = -l$.

A solution to this integral can be found first by approximating equation (2.52) by a linear relation, namely

$$E_M - E_R = c(E_t - E_R),$$

so that the sum in the denominator of the integrand in equation (2.55) is reduced to a quadratic expression, and then by solving the simplified equation (2.55) using very small increments in $V_D$ for which the correct value of $c$ is taken; finally all solutions for all increments from 0 to $V_{Dm}$ are added and the desired correct solution is obtained.

Correspondingly, the resulting increments in $T_g$, which may be called $\Delta T$, are:[8]

$$\Delta T = \int_{V_{D1}}^{V_{D2}} \frac{dE_R}{\sum\limits_{n=0}^{2} a_n E_R^n} = \frac{2l}{(a_1^2 - 4a_0 a_2)^{\frac{1}{2}}} \tanh^{-1} \frac{2a_2 E_R + a_1}{(a_1^2 - 4a_0 a_2)^{\frac{1}{2}}}\Bigg|_{E_1}^{E_2}$$

(2.56)

where for $E_M > E_s$ (equation (2.53b)),

$$a_0 = E_t^2 \mu_0 \left[ \frac{1}{2}\left(1 + \frac{\mu_0}{\mu_n}\right) - \left(\frac{3}{2}\right)k^2\frac{\mu_0}{\mu_n} - k\left(c - 1 - \frac{\mu_0}{\mu_n}\right) \right] \quad (2.57)$$

$$a_1 = E_t \mu_0 \left[ \left(1 + 2\frac{\mu_0}{\mu_n}\right) + (k-1)(c-1) \right] \quad (2.58)$$

$$a_2 = \mu_0(c - \tfrac{1}{2})$$

and for $E_t < E_M < E_s$,

$$\left.\begin{aligned}
a_0 &= E_t^2 \mu_0 \left[ \frac{1}{2}\left(1 + \frac{\mu_0}{\mu_n}\right) - \frac{1}{2}\frac{\mu_n}{\mu_0}\left(c - 1 - \frac{\mu_0}{\mu_n}\right)^2 \right] \\
a_1 &= E_t \mu_0 \left[ -\left(1 + \frac{\mu_0}{\mu_n}\right) + \frac{\mu_n}{\mu_0}\left(c - 1 - \frac{\mu_0}{\mu_n}\right)^2 \right] \\
a_2 &= \mu_0 \left[ \frac{1}{2}\left(1 + \frac{\mu_0}{\mu_n}\right) - \frac{1}{2}\frac{\mu_n}{\mu_0}\left(c - 1 - \frac{\mu_0}{\mu_n}\right)^2 \right]
\end{aligned}\right\} \quad (2.60)$$

or
$$a_0 = a_2 E_t^2 = -\frac{a_1 E_t}{2}$$

The values of $c$ are found from equation (2.52):

$$c \cong \frac{d(E_M - E_R)}{dE_R} = \left[\frac{l2en_0}{\varepsilon}\right]^{\frac{1}{2}} \frac{1}{2[E_t^* - E_R]^{\frac{1}{2}}} \quad (2.61)$$

and are given by the following table for $l = 100$ μm and $n_0 = 10^{15}/\text{cm}^3$.

TABLE 1

| $(E_t - E_R)$ (V/cm) | $c$ |
|---|---|
| 0 | $\infty$ |
| 10 | $7.5 \times 10^2$ |
| 100 | $2.5 \times 10^2$ |
| 500 | $1.2 \times 10^2$ |
| 1000 | $0.8 \times 10^2$ |

As the domain grows very rapidly for $E_t = E_R$, the singularity for $c$ at $E_t - E_R = 0$ can be neglected in connection with an assessment of domain-growth time. In particular, if, for pulse regeneration as outlined in later chapters, the bias voltage is always kept below threshold and a domain is triggered by a short input pulse, $E_R$ falls quickly below $E_t$. The domain reaches the steady state when the dotted line in Figure 2.8

has been reached. However, we assume here that steady state is reached for $E_R - \tfrac{1}{2}E_t$, which is of course only correct for large $nl$ products as the dotted line then approaches $E_R = \tfrac{1}{2}E_t$. It is obvious that the error caused by introducing $c$ can be kept very small by introducing a sufficient number of integration steps. We are interested in a solution exhibiting the general behaviour of domain transients and have to introduce suitable approximations in order to make the result more easily understandable.

With common values for Gunn-effect materials ($\mu_0/\mu_n = 3$, $k = \tfrac{1}{2}$), one finds that for $E_M > E_s$,

$$a_0 \approx -E_t^2 \mu_0 c$$

$$a_1 \approx -E_t \mu_0 c$$

$$a_2 \approx \mu_0 c$$

and for $E_t < E_M < E_s$,

$$a_0 \approx -\tfrac{1}{2}E_t^2 \mu_n c^2$$

$$a_1 \approx E_t \mu_n c^2$$

$$a_2 \approx -\tfrac{1}{2}\mu_n c^2$$

Inserting this into equation (2.56), one finds for $E_M > E_s$,

$$\Delta T_a = \frac{2l}{E_t \mu_0 c 5^{\frac{1}{2}}} \tanh^{-1} \left.\frac{2E_R - E_t}{5^{\frac{1}{2}}E_t}\right|_{V_{D1}}^{V_{D2}} \tag{2.62a}$$

and for $E_t < E_M < E_s$,

$$\Delta T_b = \frac{4l}{E_t \mu_n c^2 3^{\frac{1}{2}}} \tanh^{-1} \left.\frac{2E_t - E_R}{E_t 3^{\frac{1}{2}}}\right|_{V_{D1}}^{V_{D2}} \tag{2.62b}$$

According to equation (2.61), $c$ is proportional to $l^{\frac{1}{2}}$. $T_a$ is therefore proportional to $l^{\frac{1}{2}}$, whereas $T_b$ is independent of $l$.

In order to assess the effect of a load resistance $R_L$, applied in series to the diode, one can produce the following arguments: $R_L$ affects equation (2.52) because it has to be included in the expression for $V_D$, i.e.

$$V_D = V_B - E_R l - I_D R_L$$

where $I_D$ is the diode current during domain transit.

With $I_D = E_R \mu_0 A e n_0$, where $A$ is the cross-sectional surface of diode, this becomes

$$V_D = V_B - E_R(l + R_L \mu_0 A e n_0) \tag{2.63}$$

One sees that $R_L$ has the effect of increasing $l$ by $l_R = R_L \mu_0 A e n_0$. It follows that a series resistance increases the growth time $T_a$ by $(l + l_R)^{\frac{1}{2}}$.

Equation (2.62) can be approximated using the Taylor series approximation

$$\tanh^{-1} x \approx x \quad \text{for} \quad x^2 < 1$$

which yields from equation (2.62a),

$$T_a = \sum \Delta T_a = \frac{2l}{E_t \mu_0} \left[ \sum_{n=0}^{n=\frac{1}{2}m-1} \left( \frac{1}{m} \frac{4 - 8c_n}{41 - 2c_n - 9c_n^2} \right) \right]$$

where the integration steps are from

$$E_R = \frac{m - n}{m} E_t$$

to

$$E_R = \frac{m - n - 1}{m} E_t$$

and where $m$ is even.

$c_n$ is the value of $c$ for the integration step $n$, and is given as

$$c_n = \frac{1}{2} \left( \frac{2len_0}{\varepsilon} \right)^{\frac{1}{2}} \left[ E_t \left( 1 - \frac{m - n - 1}{m} \right) \right]^{-\frac{1}{2}}$$

The $c_n$ values of interest are larger than $10^2$ so that

$$T_a = \frac{2l}{E_t \mu_0} \left( \frac{\varepsilon}{2len_0} \right)^{\frac{1}{2}} (E_t)^{\frac{1}{2}} \sum \frac{1}{m} \left( \frac{n+1}{m} \right)^{\frac{1}{2}}$$

The expression approaches a correct solution if $m$ is going to infinity, i.e.

$$S_a = \lim_{m \to \infty} \sum_{n=0}^{n=\frac{1}{2}m-1} \frac{1}{m} \left( \frac{n+1}{m} \right)^{\frac{1}{2}}$$

This can be evaluated as follows:

$$S_a = \lim_{m \to \infty} \int_0^{\frac{1}{2}m-1} s^{\frac{1}{2}} \, ds$$

with $s = n + 1$,

$$S_a = \lim_{m \to \infty} \tfrac{2}{3} s^{\frac{3}{2}} \Big|_0^{\frac{1}{2}m-1}$$

$$= \lim_{m \to \infty} \tfrac{2}{3} (1 + n)^{\frac{3}{2}} \Big|_0^{\frac{1}{2}m-1}$$

$$= \lim_{m \to \infty} \frac{2}{3} \left[ \left( \frac{m}{2} \right)^{\frac{3}{2}} - 1 \right] = \frac{1}{3 \cdot 2^{\frac{3}{2}}}$$

Similarly one obtains a series for $T_b$,

$$S_b = \lim_{m \to \infty} \frac{1}{m^2} \sum_0^{\frac{1}{2}m-1} (n + 1)$$

which gives
$$S_b = \lim_{m \to \infty} \frac{1}{m^2} \left(\frac{m}{2}\right)^2 \cdot \frac{1}{2} = \frac{1}{8}$$

As the domain transit time
$$t_t = \frac{l}{E_R \mu_0}$$

and the dielectric relaxation time is

$$\tau_d = \frac{\varepsilon}{\mu_0 n_0 e}$$

the expressions for $T_a$ reduce to

$$T_a = \frac{(t_t \tau_d)^{\frac{1}{2}}}{3 \cdot 2^{\frac{1}{2}}} \tag{2.64a}$$

Almost the same result can be obtained by a very approximative argument, which says that the domain capacitance $C_d$ has to be charged via the low-field resistance $R_0$ so that

$$T_g \simeq R_0 C_d \tag{2.65}$$

with $R_0 = l/(ne\mu_0 A)$ and $C_d = (\varepsilon A)/d_d$, where $A$ is the cross-sectional surface of the diode and $d_d$ is the length of the domain depletion layer. $d_d$ can be found for complete depletion from Poisson's equation, i.e.

$$V_D = \frac{e d_d^2 n_0}{2\varepsilon}$$

Inserting this into equation (2.65) together with the same approximations as above, one obtains

$$T_g \simeq (t_t \tau_d)^{\frac{1}{2}}$$

Deriving the expression for $T_b$, one finds

$$T_b = \tfrac{8}{3} \tau_n S_b = \tfrac{1}{3} \tau_n \tag{2.64b}$$

where $\tau_n$ is the dielectric relaxation time of the modulus of the negative differential resistance. Under normal conditions, $E_M$ will pass $E_s$ very quickly and operate under the conditions of equation (2.64a). With reference to equation (2.15), page 15, it can be found that

$$E_s = E_R + \left[ (E_t - E_R) \frac{2 l e n_0}{\varepsilon} \right]^{\frac{1}{2}}$$

This is a quadratic equation giving the values of $E_R$ where the change-over from $E_M < E_s$ to $E_M > E_s$ occurs, for a given $l$. This shows that usually only for the very initial growth of a domain is equation (2.62a) relevant, e.g. for $l n_0 \simeq 10^{13}/cm^2$ at $E_t - E_R \simeq 10$ V/cm. Therefore equation (2.64b) gives the initial growth rate, and it is interesting that equation (2.8) presents a small-signal growth rate which is given by the same dielectric relaxation time, i.e. $\omega_d = 1/\tau_d$ for $dv/dE = \mu_n$ in this case. The $n_0 l$ condition when equation (2.64b) is entirely responsible for domain growth, can also be obtained from the above equation together with the equal-areas rule (see dotted line of Figure 2.8, page 21) which gives $E_R$ for $E_s$. One finds then that $n_0 l \simeq 5 \times 10^{10}/\text{cm}^2$, which is below the condition for domain formation, equation (2.9), so that it is unlikely that equation (2.64b) would ever be responsible for the entire growth of a domain.

As explained in the next section, $R_L$, the series load resistance to the diode, should be equal to the diode low-field resistance $R_0$ for maximum output-signal powers. This means, when using the idea of an effective increase in $l$ by $R_L$ as shown in connection with equation (2.63), the transit time $t_t$ in equation (2.64a) is effectively doubled. As $T_g$ must be twice the actual transit time of a domain, equation (2.64a) becomes

$$T_a \approx \frac{1}{3 \cdot 2^{\frac{1}{2}}} (t_t \tau_d)^{\frac{1}{2}} \approx \frac{t_t}{2}$$

which gives
$$t_t \approx \tfrac{2}{9}\tau_d$$

or
$$T_a = \tfrac{1}{9}\tau_d$$

As $\tau_d \approx \tfrac{1}{3}\tau_n$ with $\mu_0/\mu_n = 3$,

$$T_b/T_a \approx 9$$

which means that the growth time associated with equation (2.62b) is nine times larger than that of equation (2.62a).

It has to be pointed out that for small $nl$ products, the upper limit of integration of equation (2.51) is no more $E_R = \tfrac{1}{2}E_t$, but is given by the equal-areas rule, i.e. the dotted line of Figure 2.8, page 21. The above summations should therefore only be taken from $n = 0$ to some value such as $n = \tfrac{1}{4}m - 1$. The resulting sum then gives shorter growth times. This also means that a mature domain is then much reduced in amplitude and the resulting current pulses are weak, as $E_R$ is only reduced to around $\tfrac{3}{4}E_t$. This region of parameters is of no great interest for pulse-processing devices.

Computer results regarding domain transients are shown by Figures 2.9 to 2.15. Figure 2.9(a) defines two convenient time parameters in order to compare the domain growth phenomena, namely the delay time $t_d$ and the rise time $t_r$. ($E_{R0}$ and $E_{R\infty}$ are the values of field outside the domain for zero domain amplitude and fully-grown domain respectively.) The computation is based on a carrier density profile exhibiting a domain-nucleating notch as shown by Figure 2.9(c). The computation employs equation (2.51), using the $v(E)$ characteristic as given by Figure 2.8. Figure 2.9(b) shows how the delay time increases with increasing bias

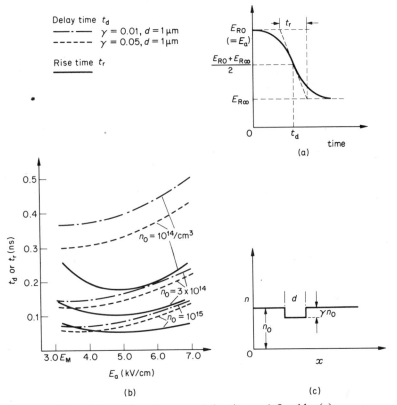

Delay time $t_d$
— · — $\gamma = 0.01$, $d = 1\,\mu m$
— — — $\gamma = 0.05$, $d = 1\,\mu m$

Rise time $t_r$
———

$E_{RO}$
$(= E_a)$

$\dfrac{E_{RO} + E_{R\infty}}{2}$

$E_{R\infty}$

time

(a)

$t_d$ or $t_r$ (ns)

0.5

0.4

0.3

0.2

0.1

$n_0 = 10^{14}/cm^3$

$n_0 = 3 \times 10^{14}$

$n_0 = 10^{15}$

3.0 $E_M$   4.0   5.0   6.0   7.0

$E_a$ (kV/cm)

(b)

$n$

$d$

$\gamma n_0$

$n_0$

$x$

(c)

**Figure 2.9**   Delay time $t_d$ and rise time $t_r$ defined by (a), for increasing bias field $E_a$.

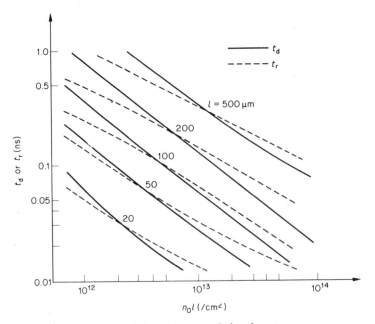

$t_d$ or $t_r$ (ns)

1.0

0.5

0.1

0.05

0.01

$t_d$
$t_r$

$l = 500\,\mu m$

200

100

50

20

$10^{12}$   $10^{13}$   $10^{14}$

$n_0 l$ (/cm²)

**Figure 2.10**   Delay time $t_d$ and rise time $t_r$ vs. $n_0$.

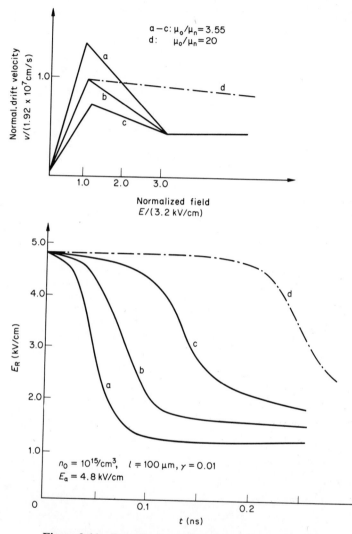

**Figure 2.11** Domain formation times for various $v(E)$ characteristics.

voltage $V_B = lE_a$, but that the rise time exhibits a minimum for $E_a \cong 5$ kV/cm. Figure 2.10 gives these two time constants *vs.* the $nl$ product for various values of $l$. The influence of temperature on transients is demonstrated by Figure 2.11, where the reduction of the peak-to-valley ratio is taken again as a three line approximation; this reduction is the same as that explained in connection with Figure 2.1, page 6. The dash-dotted line shows how the delay time very much depends on the negative differential mobility.

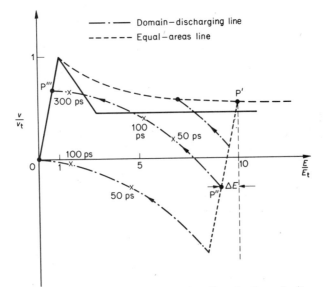

**Figure 2.12** Domain-discharging lines in the velocity-field characteristics. When a decrease in bias voltage $\Delta E\ l$ is applied for a stable domain (P′) in transit, the operating point is moved to P″, from where it moves to P‴ during the times as indicated. ($n_0 = 10^{14}/\text{cm}^3$, $L = 100\ \mu\text{m}$.)

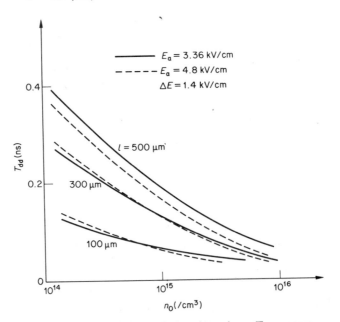

**Figure 2.13** Domain discharging time $T_{dd}$ versus ionized donar densities.

A similar computation is outlined in Figure 2.12 for the domain collapse due to a reduction in bias voltage. This figure shows the $v(E)$ characteristic and the equal-areas line. When the bias voltage is reduced by $\Delta E\,l$, the operating point moves from P′ to any of the P″ points, from where the discharging takes place again in accordance with an unequal-areas rule in the same manner as used for establishing equation (2.51). The discharging lines together with the delay times are shown in Figure 2.12. The corresponding time constants are given by Figure 2.13, and the corresponding time functions for displacement current, drift current and total current are presented by Figure 2.14. Finally, the transients for a domain running into the anode can be estimated by assuming that the domain velocity and the fields $E_M$ and $E_R$ are kept approximately unaltered. The results are given in Figure 2.15, showing basically similar behaviour to the case of Figure 2.14. All these theoretical results are partly verified by some first experimental results.[34]

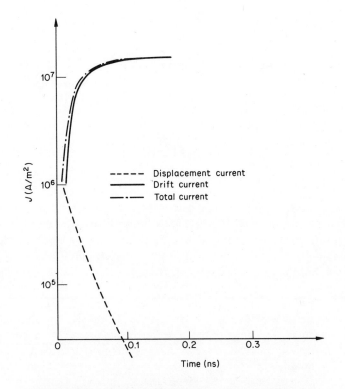

**Figure 2.14** Current density $J$ vs. time for domain-discharging process for decrease in bias voltage by $\Delta E\,l$. ($\Delta E = 2.4$ kV/cm, $E_a = 4.8$ kV/cm, $n_0 = 10^{15}/$ cm³, $l = 100$ μm.)

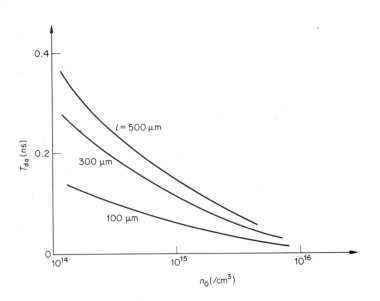

**Figure 2.15** Domain-collapsing time $T_{da}$ when running into the anode electrode versus ionized donar density $n_0$.

## 2.4   CIRCUITRY FOR GUNN-EFFECT DOMAIN DEVICES

We have seen in the previous sections that the Gunn effect exhibits itself by producing large periodic changes of current when a bias voltage is applied. This alternating current has to be supplied to a load if use is to be made of the instability. There are basically two different types of application, firstly the generation or amplification of microwave signals of relatively narrow bandwidth permitting reactive circuitry to be employed, and secondly, the possibility of generating or processing of pulse signals, where the circuit must be purely resistive in order to treat properly all the harmonics which are involved in the signal wave shape. The subject of this book concerns primarily the latter type of application, although the various modulation systems of base band pulse signals require a knowledge of the functioning of microwave circuitry for negative resistance devices such as Gunn-effect diodes. Both subjects are therefore discussed here.

### 2.4.1   *Pulse-processing circuits*

The basic problem is to avoid any distortion of the pulse signals by a dispersive network. Firstly, therefore, the system should ideally be matched

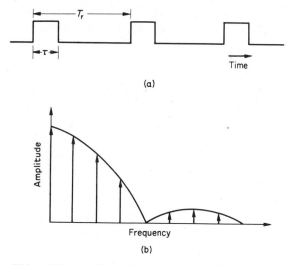

**Figure 2.16** (a) Pulse-signal wave and (b) corresponding spectrum.

everywhere. With the spectrum of a pulse wave (see Figure 2.16), the frequency range is considerable. It is useful, however, to say that the contributions by higher harmonics of the fundamental pulse-shape frequency (i.e. $f_0 = 1/(2\pi \cdot 2\tau)$, where $\tau$ is given in Figure 2.16(a)), decrease rapidly in importance with increasing harmonic number. This means that it is possible to treat approximately all parts of the circuitry which lie within a wavelength

$$\lambda_0 = \frac{c_0}{(\varepsilon_r \mu_r)^{\frac{1}{2}}} \cdot 4\pi\tau$$

(where $c_0$ is the velocity of light and $\varepsilon_r$ and $\mu_r$ are the relative permittivity and permeability), by using a d.c. consideration. This is in fact a good reason for using monolithic I.C. technology when setting up Gunn-effect logic, as the dimensions can only then be kept at a minimum. It is, of course, possible to discuss this question also in the time domain, because it only makes sense to consider a mismatch when a signal step can travel its own wavelength after reflection.

Unfortunately, a Gunn-effect diode is a two impedance device, that is, the resistance during domain transit is up to double the low-field resistance which is present when no domain exists. This property is, of course, the same as that of the two current states. It means that it is not possible to produce a circuit for these devices which is always matched, and some suitable compromise has to be made.

The second requirement in connection with a distortion-free network is that the impedance values of individual components and the transmission characteristics of a line must be independent of frequency. Therefore the use of reactive components, such as capacitors, inductors, and transformers, has to be excluded. Similarly, dispersive transmission lines such as waveguides are not suitable, and either coaxial or microstrip lines have generally been employed. It is, of course, sometimes possible to use subsequent compensation networks after some frequency distortion has been introduced. Unfortunately they introduce delays which cannot be tolerated when complex pulse-processing networks are considered. For simple pulse regeneration, however, it can be attractive to employ such compensation systems as long as no important parts of the pulse spectrum have disappeared below the noise level. Pulse regenerators do not rely on fast speeds of pulse processing but only on correctly reshaping and retiming of the pulse signals employed, even if they are extremely short as is envisaged nowadays.

The best method of extracting signal power from these domain devices is therefore to connect them in series with a load resistance. In fact, as the devices act as pulse current sources, the load resistance $R_L$ should be as large as possible if a maximum signal voltage is required. However, for increasing $R_L$ the load line (see Figure 2.17), which can be made to include the effect of $R_L$ as explained on page 39, becomes very steep so that $V_D$ will assume large values. When the domain voltage grows too high, the maximum field $E_M$ can exceed the avalanching field and electron-hole pairs are generated. Their life-times are unfortunately much longer than the domain transit time, so that they accumulate to such high values

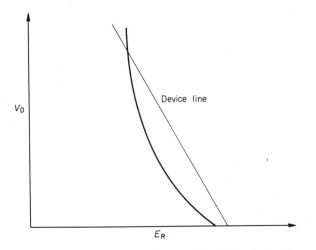

**Figure 2.17** Domain characteristics. The device line is given by $l_1 = l + R_L \mu_0 e n_0 A$.

after several domain passages that the electron density becomes too large for the electron-transfer effect to occur, and the Gunn-effect device represents a short circuit.

The effect of $R_L$ on the device line of Figure 2.17 can be shown by using the same argument as above in connection with equation (2.63). There, the load resistance $R_L$ was shown to increase the effective length of the diode so that $l$ has to be replaced by $l_1 = l + R_L \mu_0 e n_0 A$. The slope of the device line is now given by $l_1$ and this becomes steeper with increasing· $R_L$.

Usually, however, it is not the maximum pulse-signal voltage, but pulse-signal power which has to be transferred into a transmission line of a given characteristic impedance $Z_L$. It is then useful to employ a load resistance which is equal to $Z_L$. In fact, if a suitable current path even for the direct current via $Z_L$ can be arranged, no separate load resistor is needed as $Z_L$ can act as series load. It is emphasized again, however, that owing to the pulse spectrum of Figure 2.16(b) the d.c. path must also go via a load of $Z_L$. This excludes low-impedance d.c. connections. In order to have maximum signal power, the domain device should therefore be as small as possible so that a maximum amount of current is switched by it. If, however, the diode resistance becomes much smaller than $R_L$, the signal will not be fully transmitted into the line, but will partly be shortened via the diode. Also, if an incoming signal is to nucleate a domain, this signal cannot then be coupled effectively to the diode owing to the mismatch with the input transmission line. The diode should thus be of the same order of magnitude as the transmission line impedance. This determines, therefore, the maximum signal power $P_S$ which can be delivered into a transmission line, namely

$$P_S = \frac{(1-k)^2 E_t^2}{2Z_L} l^2$$

For room-temperature operation ($k \simeq 0.5$, see Figures 2.1 and 2.4, pages 6 and 14) and for a common characteristic impedance $Z_L = 50\ \Omega$, the values of $P_S$ for $l = 100\ \mu m$ and $10\ \mu m$ are 2 W and 20 mW respectively. Unfortunately, the devices have to be operated with a power dissipation requirement of

$$P_d = \left(\frac{E_t l}{R_0}\right)^2$$

where the worst case of no signal occurrence is considered. $P_d$ is 18 W and 180 mW for $l$ equal to 100 $\mu m$ and 10 $\mu m$ respectively. This is caused by a disadvantage of Gunn-effect pulse devices, namely that no 'normally off' operation can be arranged. As will be shown later (section 4.5), the minimum thermal resistances $\theta_T$ achievable for pulse-processing devices result in relatively high operating temperatures of the devices ($\theta_T$ is the ratio of heat-power flow to maximum temperature increase for the active device parts), and operating temperatures of around 100 to 200 °C have to be accepted for bias voltages near threshold. For such elevated temperatures $k$ increases and takes a value of $\frac{2}{3}$ for good d.c.-biased devices. Therefore the corresponding signal powers given above are reduced by the coefficient $\frac{4}{9}$.

With $R_L$ commonly connected in series with the diode, the transmission line can either be applied in parallel to $R_L$ or to the diode itself. In fact, sometimes the input transmission line is connected parallel to the diode, whereas the output line is supplied from $R_L$ or *vice versa*. Then the input signal has a different polarity from the output pulse. Generally, however, one can say that a resistor in parallel to the diode (in addition to the finite series load) has the result of effectively reducing the domain-pulse amplitude, as part of the current through the series load is supplied via this passive parallel resistor. Therefore it can be concluded that none of these alternatives have any particular advantage from the signal power point of view.

A serious disadvantage of any two-terminal Gunn-effect device is that domain nucleation can be achieved with signals arriving at the transferred-electron diode via either the input or the output line. In fact there is no criterion which allows one to distinguish between output and input connections. From this point of view there is thus no advantage in the possibility mentioned above of arranging input and output lines such that signals of opposite polarities would travel along them. Indeed, if a domain is nucleated by a pulse arriving along the output line, a new output pulse is launched into the input line in the reverse direction. The result is that a reflected signal can travel back to any diode and produce spurious domains. Therefore either fast, non-dispersive, unidirectional elements have to be inserted near the output terminal, or a structure has to be developed which is a three-terminal Gunn-effect system directly exhibiting directionality. Both methods have been considered with varying success, and further details will be found in Chapter 3.

For an application of the Gunn effect domain for ultrafast pulse-signal processing, there are two parameters of importance, firstly, the input power $P_{im}$ required to trigger domains reliably, and secondly the time it takes domains to grow to full size. The minimum input power is most probably related to the minimum time $\tau_m$ required for this power to be applied, where $\tau_m$ would be related to the diode $n$ and $l$. The product of $P_{im} \times \tau_m$ is the realistic counterpart for the often-employed product of power $\times$ switching time for passive switching devices. Although there have been several attempts to obtain the $P_{im} \times \tau_m$ product theoretically and experimentally, no realistic data is available yet. The theoretical approaches have generally used Johnson noise as the relevant phenomenon being responsible for causing the limits of the input signal parameters. Experimental methods employ either very short input pulses with reducing amplitude—a technique which is open to errors due to the dispersive nature of any mounting at the extreme frequencies required for this test—or a microwave signal of increasing frequency and decreasing amplitude where the input signals do not require any more the extreme non-dispersive requirements for the mounting. However, owing to the technological difficulties of this experiment, no reliable results have been obtained yet, nor has any agreement with theory been evolved. It can, however, be said that the minimum input energy required for a diode with $R_0 = 50\ \Omega$ and $l = 100\ \mu m$ is unlikely to be higher than $10^{-2}$ pJ. This is very much smaller than the product switching time $\times$ power for classical devices.

The other parameter of importance is the time it takes a domain to grow to full maturity. This information can be obtained with the theory of the last section, particularly the equal-areas rule for transients or the unequal-areas rule. This then depends strongly on the $nl$ product and $\mu_n$. As an example, a device with $l = 100\ \mu m$ with an $ln = 10^{13}/cm^2$ switches fully within 70 ps if operated under pulsed bias so that the device temperature is around 30 °C, and within 200 ps if d.c. operated when the device temperature is rather elevated. These are experimental results obtained so far, which can be improved further by better circuitry and fabrication technology.

### 2.4.2 *Microwave narrow-band circuits*

In order to generate or process microwave signals, both Gunn effect devices and avalanche diodes can be operated in a suitable resonant circuit.[6,20] The active element can be represented by an equivalent circuit. Its component values depend strongly, on the type of device mode. For example, whether Gunn-effect domains are quenched by the r.f. voltage swing caused by the resonant circuit, or whether they can run into the anode contact where they then collapse due to the high $n$ there, has a very strong effect on the values of the equivalent circuit. This equivalent circuit has been calculated for a few modes, sometimes with severe approximations in order to make the mathematical problem manageable.

Typical results have been obtained for the Gunn-effect quenched-domain mode by approximating the current-voltage relation under transient conditions including domain growth and decay times, where (a) partly experimental data such as the subthreshold $I/V$ behaviour, (b) partly theoretical predictions such as the $V_D$ *vs.* $E_R$ dependence, and (c) partly purely phenomenological descriptions such as exponential expressions for domain transients, have been used.[42] This treatment is therefore more trustworthy than the more simplified method which assumes instantaneous domain formation and decay (e.g. reference 1). For a purely

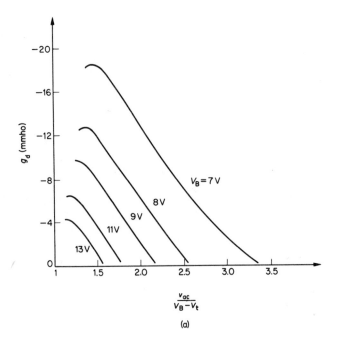

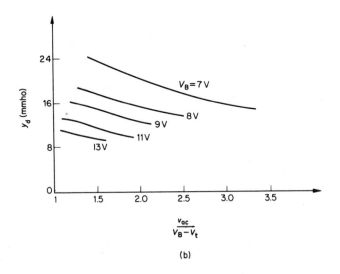

**Figure 2.18**  Calculated negative conductance $g_d$ and susceptance $y_d$ *vs.* normalized microwave voltage $v_{ac}$ for various bias voltages $V_B$ of Gunn-effect diodes.

sinusoidal voltage one obtains a sophisticated current waveform, which is then Fourier analysed. The fundamental of this current wave is out of phase with the voltage wave. The ratio of the r.f. current to r.f. voltage at this fundamental frequency yields the complex device admittance, whose conductive part is negative and whose susceptance is capacitive. The two terms are shown in Figure 2.18 as a function of normalized r.f. voltage $v_{ac}$ for a frequency of 10 GHz. One can see that the negative conductance decreases with increasing r.f. voltage until it becomes positive, when the device changes from an active to a passive component. Some of the features of this figure are, of course, determined by the model employed. For example, for very large bias voltages the negative conductance $g_D$ is shown to become zero. It is likely that actual devices will not exhibit such a behaviour as a different mode of operation will probably take over.

The device admittance $g_d + j\omega C_d$ is usually transformed to $g'_d + j\omega C'_d$ due to the package reactances whose equivalent circuit has been assessed in detail.[46] Often it is sufficient to represent this by the equivalent circuit of Figure 2.19(a), which is also quite instructive as the different reactances

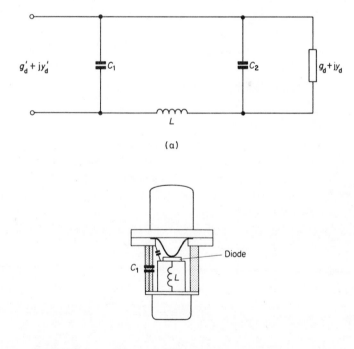

(a)

(b)

**Figure 2.19** (a) Equivalent circuit of diode and package, (b) Sketch of diode in S4 package with individual parts of equivalent circuit indicated.

*Gunn-effect Logic Devices*

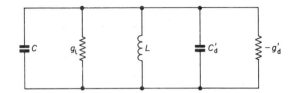

**Figure 2.20** Equivalent circuit of diode and resonator ($C$, $L$ and $g_1$ are the capacitance, inductance, and load conductance representing the resonator, $C_d'$ and $g_d'$ represent the encapsulated diode).

can be understood in terms of various physical parts of the package (Figure 2.19(b)).

The r.f. network of diode plus package has to be matched to the resonator impedance, as given at the point where the encapsulated diode is positioned, or, in other words, the sums of the conductive and reactive parts of all impedances must be zero. Representing the resonator by a parallel circuit as shown by Figure 2.20, this means that

$$g_1 = |g_d'| \tag{2.66a}$$

and
$$\omega C - \frac{1}{\omega L} = \omega C_d' \tag{2.66b}$$

Equation (2.66a) determines then the amplitude of the r.f. voltage across the diode in accordance with Figure 2.18(a), and equation (2.66b) fixes the frequency of operation which is, of course, also slightly dependent on $v_{ac}$ due to the slopes of Figure 2.18(b).

Experimentally, the negative device conductance can be obtained by one of the following three methods. Such information can firstly be found by measuring the impedance of the microwave network which was applied as a load to the active oscillating device and by taking the negative values of the resulting impedance as that of the device. In this way a limited range of $v_{ac}$ can be covered, as long as the slope of $g_d(v_{ac})$ is negative.

Secondly, a transient method can be employed which gives also conductance details for positive slopes.[41]

A suitable cavity has to be used. The bias voltage $V_B$ is applied as a steep step function, and the resulting oscillatory waveform gives the diode r.f. impedance as a function of $v_{ac}$. The resulting information can also be used to estimate the diode space-charge-wave mode for a given $v_{ac}$ and $V_B$. Of course, it is essential that the cavity mode remains the same during the transient, and this has to be ascertained by probing the microwave field along the resonator during the transient.

The power $P_g$ generated by the active device is equal[53] to the output power $P_L$ delivered to the load, the losses $P_l$ of the cavity, and the change in energy, $W$, stored in the cavity, i.e.

$$P_g = P_L + P_1 + \frac{dW}{dt}$$

Rewriting this equation gives

$$\tfrac{1}{2} g_d \left(\frac{v_{ac}}{v_{01}}\right)^2 = \frac{1}{2} \frac{Q_e}{Q_1} g_1 + \frac{Q_e g_1}{\omega_0} \frac{d}{dt} \log v_{01} \qquad (2.67)$$

where $g_d$ is the equivalent r.f. conductance of the device for a sinusoidal voltage swing,

$v_{ac}$ is the amplitude of the r.f. voltage across the device,

$v_{01}$ is the amplitude of the r.f. voltage across the load impedance,

$Q_e$ is the external $Q$ of the cavity,

$Q_1$ is the loaded $Q$ of the cavity,

$g_1$ is the load conductance,

$\omega_0$ is the angular frequency of operation,

and where the following conditions have to be satisfied:

1. there is no change of cavity mode, and

2. the growth time is much larger than the oscillation period.

$v_{ac}$ and $v_{01}$ are related to each other by a coupling coefficient $n_c$, i.e.

$$n_c v_{ac} = v_{01}$$

If the change of frequency is small during the build-up of oscillation, the values of $Q_e$, $Q_1$, and $n_c$ are constant and equal to the value of the steady state, when the last term of equation (2.67) vanishes. It is then possible to obtain $(Q_e/Q_1)g_1$, $n_c$, and the device's negative conductance $g_d$ for steady-state operation from steady-state measurements. With these values, the function $g_d(v_{ac})$ can be derived with the help of equation (2.67) from the measured function $v_{01}(t)$ under transient conditions.

A microwave resonator can be understood as a transmission line of a certain characteristic impedance which is shortened at one end. A Smith Chart can be employed to determine the impedance at the plane where the diode is inserted. For oscillations to occur, the transformed complex diode-impedance must show such values that the sum of the impedances is zero again. In the cavity there will then be a standing-wave pattern of field which, if measured very carefully, is an indication of the transformed diode impedance. Thus by carefully measuring the shape of the wave pattern using a loosely coupled capacitive probe which is movable along the length of the transmission line, one can also determine the diode impedance. By making the short at the opposite end from the diode movable by a sliding arrangement and by applying a coupling probe which extracts power from the cavity near the short, the transmission

*Gunn-effect Logic Devices*

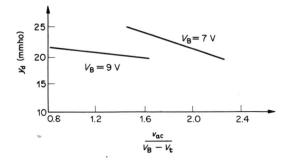

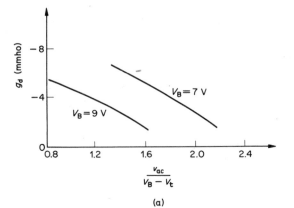

(a)

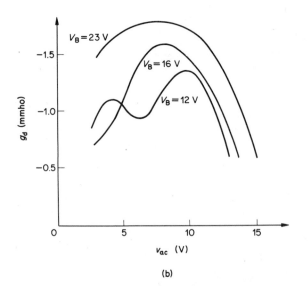

(b)

**Figure 2.21** Typical experimental admittance for domain mode Gunn-effect oscillator: (a) method based on steady-state measurement, frequency 9.7 GHz; (b) results with transient method, frequency 10.45 GHz.

line impedance at the plane of the diode can be varied. In this way the diode admittance can be measured[42] as a function of $v_{ac}$ and $\omega$ again. Typical results obtained by this method and by the transient approach are shown in Figure 2.21.

Avalanching junction devices can also exhibit a negative conductance if an alternating voltage generates bunches of excess charges near its peak values. These bunches then drift under the action of an electric field with saturation velocities across the junction depletion and, if specially provided, across an intrinsic, high-resistivity layer. During the transit of these excess charges, the terminal device current maintains a high value although the alternating voltage swings to negative values. Therefore a negative conductance effect is found, together with a capacitive behaviour.

Both experimental and theoretical results have been obtained regarding an avalanche-diode equivalent circuit consisting of a parallel negative conductance and a susceptance as given by Figure 2.22. One can see that the behaviour is similar to that of Gunn-effect diodes. Figure 2.22 shows that the susceptance of an avalanche diode tends to increase slightly with $v_{ac}$ in contrast to that of a Gunn-effect device (see Figure 2.21), whereas the conductance behaviour is very similar, except that $|g_d|$ of avalanche diodes are generally smaller than that of Gunn-effect devices. The theoretical admittance functions were found[51] by solving Poisson's equation and the continuity equation, by computing the resulting current wave for a diode imbedded in a resonant circuit, and by Fourier analysis of the terminal current and applied voltage waveforms. The results, which are obtained by taking the ratio of the current and voltage vector components at the operating frequency, again represent a large-signal case similarly to the Gunn-effect diode approach mentioned above.

## 2.5 DOMAIN LIMITATION BY IMPACT IONIZATION

When the magnitude of the electric field in the high-field domain exceeds a critical level, some electrons in the domain obtain enough energy to ionize the lattice atoms when they collide with it, thus creating a new electron-hole pair. The excess holes produced by the ionization process are quickly trapped with a life-time of less than 1 ns. Since the holes are trapped, the excess electrons generated will be localized at the hole trap sites due to space charge neutrality requirements, and the result is an apparent increase in the number of conduction electrons in the region where impact ionization occurred.[3]

If domain fields are too high, impact ionization occurs during each domain transit, thus increasing the valley current in each successive cycle, until domain formation becomes non-coherent.

From Figure 2.5 (page 15) one can see that for a 10 per cent increase in $n_0$, $E_R$ only increases by less than 0.5 per cent for a relatively high value

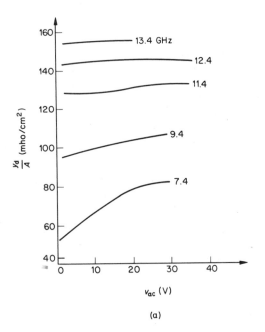

(a)

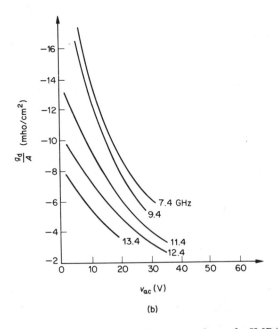

(b)

**Figure 2.22** Computational results of IMPATT diode; current density 200 A/cm²; (a) susceptance $y_d$, (b) conductance $g_d$.

of $V_D$. One can therefore say that $E_R$ is unaffected by a small increase in carrier density due to avalanching and the valley current, which is given as

$$J_v = en\mu_0 E_R$$

is proportional to $n$. By applying a pulsed bias voltage in the form of a two-step function, normal domain transits can occur first during the lower step, before the ionizing voltage level is reached with the upper step. In this way the devices could first operate in a pre-ionization mode for a short while, before ionization was initiated and results are more repeatable. By defining the ionization threshold voltage $V_{Dth}$ as that excess domain voltage which produces a 1 per cent increase in valley current over ten domain transits, an ionization threshold of between 65 and 290 V for carrier densities ranging from $10^{15}/cm^3$ to $3 \times 10^{14}/cm^3$ respectively was found. Taking the bias voltage threshold for ionization, $V_A$, as a multiple of the threshold voltage for domain formation, $V_A = MV_t = ME_t l$ with $E_t = 3.2$ kV/cm, one finds the results of Figure 2.23.

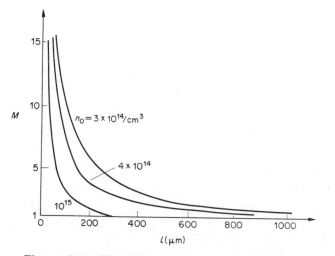

**Figure 2.23** Threshold of impact ionization for multiples of threshold voltage as a function of $l$ for various doping levels.

This figure shows very important limiting parameters for Gunn-effect pulse devices. The maximum device length for $10^{15}/cm^3$ is around 375 μm and for $4 \times 10^{14}/cm^3$ is around 1000 μm, since at these lengths impact ionization commences immediately at oscillation threshold.

The values of $V_{Dth}$ obtained experimentally so far, seem to obey a relation of the following form:[3]

$$V_{Dth} n_0^{\frac{5}{4}} = 3.6 \times 10^{20} \quad [V/cm^{\frac{15}{4}}]$$

although no theoretical significance can be applied to it, and it is only relevant for $n_0$ ranges from $10^{14}/cm^3$ to $3 \times 10^{15}/cm^3$. Although some theoretical attempts seem to support the experimental values described here, it must of course be pointed out that these are not necessarily limiting values and that some future materials of better quality might exhibit higher threshold values for $V_{Dth}$.

In connection with impact ionization, localized ionization can also occur wherever high electric fields exist, because ionization phenomena are related to a threshold-field value. Usually, high electric fields occur at the anode contact whenever a domain collapses there. Such a domain can run into the electrode faster than a dielectric relaxation time, which would be the time constant required for a re-arrangement of the fields outside the domain. During this time constant the excess-domain voltage $V_D$ of the freely-travelling domain has to be accommodated across a domain whose depletion layer has run into the $n^+$ region of the anode, and increased domain fields result. These anode fields cause local ionization, which also tends to affect seriously the output signals from Gunn-effect regenerators. In fact, the domain pulse often becomes shorter and less clear in shape towards its end. A remedy for this effect is an enlargement of the cross-sectional surface of Gunn-effect devices near the anode electrode. This is, of course, often advisable also from two further points of view.

The anode contact metals must not experience a high electric field. Therefore a widening of the anode edge is advisable unless regrown $n^+$ contacts are used. Secondly there is some evidence that a field-enhanced trapping effect due to preferential trapping of upper-valley electrons occurs.[59,61] In order to avoid any distortion of the output pulse waves from Gunn-effect regenerators by such trapping, a tapered structure with the anode being the wider part has been employed.[59] Large-signal computer results indicate that the density of trapped electrons is largest in the anode region. Too large an amount of trapping there could lower the cathode field below threshold so that no coherent domain nucleation is possible any more.

It is finally interesting to add that an ionization coefficient of the following form[65] has resulted in some good agreements between experimental results and theory for the carrier density ranges of interest here;

$$\alpha(E) = 10^6 \exp \frac{-1.72 \times 10^6}{E[kV/cm]}$$

## 2.6 STATIONARY GUNN-EFFECT DOMAINS

So far, only travelling dipole domains have been described for logic applications. It has been established, however, that under certain conditions a dipole domain can be transformed into an accumulation layer domain on arrival at the anode and can stay there in a stable equilibrium. The field distribution is then as given in Figure 2.24. The doping fluctuations and other inhomogeneities must be sufficiently small so that the field between stationary domain and cathode stays below the threshold for domain nucleation. It is therefore unlikely that stationary

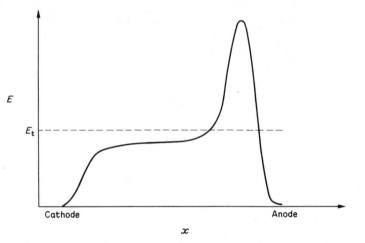

**Figure 2.24**  Field distribution of a stationary domain.

domains will easily be observed with very long interelectrode distances, and the best conditions are a continuous small increase of resistivity when going from cathode to anode.

The stationary domain is possible because more electrons flow into it from the cathode side, where the fields $E_R$ are below $E_t$ and the velocities can be high (see Figure 2.1, page 6), than leave it at the anode side, where the electrons are in satellite valleys and therefore slow. Because only an accumulation layer is involved, the domain width is very much reduced. Experimental results have been obtained with devices of 10 $\mu$m interelectrode distance, when the accumulation layer of a bipolar domain would be about 3 $\mu$m wide as given by diffusion, and the depletion layer would exhibit the same width. It can therefore be expected that the fields in the monopolar domain are twice as high as in a bipolar one for a given bias voltage. Correspondingly, it can be expected that avalanching sets in at about half the bias voltage for such a stationary domain as compared with a travelling one, because the avalanching threshold is constant as outlined in the previous section. In fact, this phenomena can be observed experimentally so that a certain verification of the concepts described here exists.

Stationary domains can be nucleated by the application of a bias voltage above the threshold for the transferred-electron effect. This voltage can remain of high value as long as the stationary domain exists. The stationary domain causes a low-current state similarly to a bipolar domain; however, the reduction in current is found to be smaller than that given by the ratio $k$, as defined by Figure 2.4, because the domain electrons discharging into the anode contact are travelling into the contact region where low mobilities and high carrier densities exist, so that no simple equal-areas rule is applicable.

A domain can be extinguished by applying a short, negative voltage pulse, bringing the domain field below the transferred-electron threshold for a sufficient time. This is therefore a bistable element which is similar to conventional bistable logic systems. The switching-on time will be as fast as with bipolar-domain devices, and the switching-off time can also be fast as the electron transfer mechanism occurs within a few picoseconds only and the discharging of the accumulation layer into the anode contact takes place within about 30 ps. This effect is therefore very suitable for logic memory applications and other fields where bistable devices are of advantage, and it can be considered to be an element which would supplement the monostable logic mainly considered in this book.

# 3: Domain and Semiconductor Surfaces

Although the Gunn effect is a volume phenomenon which only occurs in the bulk of the semiconductor, surfaces can affect the domains very seriously. This imposes important limitations on device structures, particularly when planar photolithographically-produced structures are considered. It also means that signal extraction with reasonable power levels is difficult when attempted by surface probes. This has been done on many occasions in the past in order to analyse domain fields, always, however, by using high-impedance probes, and no appreciable electronic signal powers have been extracted yet via the surfaces. This does not mean that there might not be a future development which employs some suitable surface loading method to overcome this difficulty. All these considerations demonstrate that surfaces are very important for Gunn-effect pulse-processing elements, and this chapter discusses the relevant implications.

When we employ the word surface here, then we mean all surfaces of the semiconductor element except those which carry current via ohmic contacts. Often, however, we restrict ourselves only to the free surfaces between the ohmic electrodes.

## 3.1 TRANSFERRED-ELECTRON SPACE-CHARGE WAVES AND SURFACES

When the active layer of a planar Gunn-effect element becomes too thin, or when dielectric or magnetic materials are deposited on the free surfaces, an interaction of the space-charge waves with the outside medium can reduce their amplitudes by such an amount that they might even be annihilated and cannot grow into a mature domain. The treatment investigating these phenomena represents a small-signal analysis which is, however, also applicable to domain formation as any large-signal instability always has to start first as a small-signal one.

The effective mobility of electrons in the direction of d.c. drift (i.e. the $z$-direction), is $\mu_z = \partial v / \partial E$, and the effective mobility in transverse direction (i.e. $y$-direction) is $\mu_y = v/E = \mu_d$, ($v$ = electronic drift velocity, $E$ = electric field). The static electron density is approximated to be uniform, all r.f. (radio frequency) terms vary with $\exp j(\omega t - \beta z)$, and diffusion is neglected. We consider a thin semiconducting layer of infinite extent on the $x$- and $z$-directions which is sandwiched in between another material of either high $\varepsilon_r$ or $\mu_r$ ($\varepsilon_r$ and $\mu_r$ are the relative permittivity and permeability respectively). It can be shown that only the following field components exist: $E_y$, $E_z$, and $H_x$. The small-signal Maxwell equations are then

$$\frac{\partial \tilde{E}_z}{\partial y} + j\tilde{E}_y\beta = -j\tilde{B}_x\omega \tag{3.1}$$

$$-j\tilde{H}_x\beta = n_0 e\mu_\mathrm{d}\tilde{E}_y + j\varepsilon\tilde{E}_y\omega \tag{3.2}$$

$$-\frac{\partial \tilde{H}_x}{\partial y} = n_0 e\mu_z\tilde{E}_z + j\varepsilon\tilde{E}_y\omega + \tilde{n}e\mu_\mathrm{d}E_0 \tag{3.3}$$

where $\tilde{H}$ and $\tilde{B}$ are the magnetic field and magnetic induction respectively, $n_0$ is the d.c. carrier density, $e$ is the electronic charge, $\mu_\mathrm{d}$ is the d.c. mobility and $\mu_\mathrm{d}E_0 = v_\mathrm{d}$. A tilde describes the r.f. terms and the subscript 0 the d.c. components, in the manner of the following expression:

$$E_z = E_0 + \tilde{E}_z \exp j(\omega t - \beta z)$$

Together with Poisson's equation,

$$-j\tilde{E}_z\beta + \frac{\partial \tilde{E}_y}{\partial y} = \frac{\tilde{n}e}{\varepsilon} \tag{3.4}$$

equations (3.1) to (3.3) give the solution

$$\tilde{E}_y = A_\mathrm{E} \sin \alpha y \tag{3.5}$$

where $\quad -\alpha^2 = \dfrac{[\beta_{cz} + j(\beta_\mathrm{e} - \beta)](\beta^2 + j(v_\mathrm{d}^2/c_0^2)\beta_{cy}\beta_\mathrm{e} - (v_\mathrm{d}^2/c_0^2)\beta_\mathrm{e}^2)}{\beta_{cy} + j(\beta_\mathrm{e} - \beta)} \tag{3.6}$

with $\quad \beta_{cz} = \dfrac{\mu_z n_0 e}{\varepsilon v_\mathrm{d}} \qquad \beta_\mathrm{e} = \dfrac{\omega}{v_\mathrm{d}} \qquad \beta_{cy} = \dfrac{\mu_y n_0 e}{\varepsilon v_\mathrm{d}}$

and $\qquad\qquad\qquad c_0 = \dfrac{1}{(\varepsilon\mu)^{\frac{1}{2}}}$

The expression for the Hahn boundary condition enables one to consider the interface between the semiconducting layer and the adjoining material at $y = \pm a$, i.e.

$$\varepsilon_\mathrm{II}\tilde{E}_{y\mathrm{II}} = \varepsilon_\mathrm{I}\tilde{E}_{y\mathrm{I}}\left(1 - \frac{j\beta_{cy}}{(\beta_\mathrm{e} - \beta)}\right) \tag{3.7}$$

where the subscripts I and II denote quantities of the semiconductor and adjoining material respectively. Solving the small-signal Maxwell equation for II gives

$$(-\omega^2\varepsilon\mu + \beta^2)\tilde{E}_z = \frac{\partial^2 \tilde{E}_z}{\partial y^2}$$

With $k_w = \omega/c_0$, one finds

$$\tilde{E}_z = B_E \exp \pm(\beta^2 - k_w^2)^{\frac{1}{2}}y \qquad (3.8)$$

where the sign in front of the root has to be chosen such that $\tilde{E}_z$ goes to zero for $y = \pm\infty$. At $y = \pm a$, we have the boundary condition

$$\tilde{E}_{zI} = \tilde{E}_{zII} \qquad (3.9)$$

where $\tilde{E}_{zI}$ can be obtained from (3.1) to (3.5). This gives

$$B_E \exp(\beta^2 - k_w^2)^{\frac{1}{2}}a = \frac{A_E}{\alpha\beta}\left(j\beta^2 - \frac{v_d^2}{c_I^2}\beta_e\beta_{cy} - j\frac{v_d^2}{c_I^2}\beta_e^2\right)\cos\alpha a \qquad (3.10)$$

Using equation (3.7) together with (3.5), (3.8) and (3.10), one obtains

$$\alpha \tan \alpha a = \frac{(\beta^2 - k_w^2)^{\frac{1}{2}}(\beta^2 + j(v_d^2/c_I^2)\beta_{cy}\beta_e - (v_d^2/c_I^2)\beta_e^2)(\varepsilon_{II}/\varepsilon_I)(\beta_e - \beta)}{(\beta^2 - (v_d^2/c_{II}^2)\beta_e^2)(\beta_e - \beta - j\beta_{cy})} \qquad (3.11)$$

One can show with the help of the final solution which we are going to obtain, that $|\alpha a| \ll 1$, and $\beta_e \simeq \beta$, as long as surface loading is only light (e.g. small $\varepsilon_{II}/\varepsilon_I$).

Using these approximations together with equations (3.6) and (3.11), one finds

$$\beta = \beta_e - j\beta_{cz}\frac{\beta_e a(1 - v_d^2/c_{II}^2)^{\frac{1}{2}}}{\varepsilon_{II}/\varepsilon_I + \beta_e a(1 - v_d^2/c_{II}^2)^{\frac{1}{2}}} \qquad (3.12)$$

As the solution for large $a$ is

$$\beta = \beta_e - j\beta_{cz} \qquad (3.13)$$

where $\beta_{cz}$ becomes negative above the Gunn-effect threshold, thus showing wave growth, equation (3.12) indicates how the growth rate is reduced by the factor

$$\frac{\beta_e a}{(\varepsilon_{II}/\varepsilon_I)[1 - (v_d^2/c_{II}^2)]^{-\frac{1}{2}} + \beta_e a} \qquad (3.14)$$

This shows that both dielectric and magnetic materials along the surface of GaAs can inhibit domain formation.

As in all our cases,

$$\beta_e a \ll \frac{\varepsilon_{II}}{\varepsilon_I}\left[1 - \frac{v_d^2}{c_{II}^2}\right]^{-\frac{1}{2}} \qquad (3.15)$$

we can determine the condition for domain formation in the following manner.

As indicated in Chapter 2 by equation (2.9), Gunn-effect domains can be formed when the condition $nl > 10^{11}/\text{cm}^2$ is satisfied. According to equation (3.13), this corresponds to an r.f.-field growth by a factor

$\exp |\beta_{cz} l|$ in the length of the diode where

$$|\beta_{cz} l| = |\omega_{cz} l / v_d| = nl/(3 \times 10^3 v_d) \tag{3.16}$$

Taking $nl \sim 5 \times 10^{11}/\text{cm}^2$, we find that the condition becomes $\beta_{cz} l \geqslant 1.7 \times 10^8/v_d$.

With the inequality of (3.15) one finds from equation (3.12) that

$$\beta a \left[ 1 + j\beta_{cz} a \frac{\varepsilon_I}{\varepsilon_{II}} \left( 1 - \frac{v_d^2}{c_{II}^2} \right)^{\frac{1}{2}} \right] = \beta_e a$$

or

$$\beta a = \frac{\beta_e a [1 - j\beta_{cz} a (\varepsilon_I/\varepsilon_{II})(1 - v_d^2/c_{II}^2)^{\frac{1}{2}}]}{1 + |\beta_{cz} a|^2 (\varepsilon_I/\varepsilon_{II})^2 (1 - v_d^2/c_{II}^2)} \tag{3.17}$$

For a thin film, this expression indicates a growth factor of

$$\text{Im } \beta l = \frac{\beta_{cz} a (\varepsilon_I/\varepsilon_{II})(1 - v_d^2/c_{II}^2)^{\frac{1}{2}}}{1 + |\beta_{cz} a|^2 (\varepsilon_I/\varepsilon_{II})^2 (1 - v_d^2/c_{II}^2)} \beta_e l \tag{3.18}$$

For oscillations to occur, the condition

$$\text{Re } \beta l = \frac{\beta_e l}{1 + |\beta_{cz} a|^2 (\varepsilon_I/\varepsilon_{II})^2 (1 - v_d^2/c_{II}^2)} \approx 2\pi \tag{3.18a}$$

must be satisfied. From both conditions (3.18) and (3.18a) it follows that the criterion for domain formation is

$$2\pi\beta_{cz} a \frac{\varepsilon_I}{\varepsilon_{II}} \left( 1 - \frac{v_d^2}{c_{II}^2} \right)^{\frac{1}{2}} \geqslant \frac{1.7 \times 10^9}{v_0}$$

With the thickness $d = 2a$ and $n$, the equilibrium carrier density, we obtain the minimum $nd$ product, i.e.

$$nd \geqslant 1.6 \times 10^{11} \frac{\varepsilon_{II}}{\varepsilon_I (1 - v_d^2/c_{II}^2)^{\frac{1}{2}}} \text{ cm}^2 \tag{3.19}$$

This expression is only valid of course for light loading, i.e.

$$\frac{\varepsilon_{II}}{\varepsilon_I (1 - v_d^2/c_{II}^2)^{\frac{1}{2}}} \tag{3.19a}$$

is not too large as otherwise the approximation $\alpha \tan \alpha\, a \simeq \alpha^2 a$ does not hold any more. In order to find also the $nd$ product for heavy loading, a numerical analysis has to be performed of the dispersion equation. The correct treatment of dispersion equations can only be performed by the technique of mapping the complex $\beta_e$ values into the complex $\beta$-plane and of searching for saddle points, where a double root of $\beta$ for some complex $\beta_e$ with a negative imaginary part $\beta_{ei}$ of $\beta_e$ occurs (see Chapter 3 of reference 20). When the two merging roots originate from different upper and lower halves of the complex $\beta$ plane for a large negative $\beta_{ei}$, one has a non-convective or absolute instability. Such a saddle point means then that a local perturbation grows at this point in space. On the other hand, a convective instability can also occur, which is a growing perturbation travelling with the drifting charge carriers. This is exhibited by the mapping technique if $\beta_i$, the imaginary part of $\beta$, has a different sign when the frequency $\beta_e$ takes on a large negative imaginary part.

Such an analysis has been performed,[31] and an absolute instability was found for the values of (3.19a). When the imaginary part of the saddle frequency, $\beta_{esi}$, becomes positive, the wave stabilizes, and this occurs below a certain $nd$ product. It is therefore possible to establish for this absolute instability a critical $nd$ product by taking $\beta_{esi} = 0$ as a function of the expression (3.19). Figure 3.1 shows a typical result for $\mu_d = 2600$ cm²/V s and $\mu_z = -2000$ cm²/V s (corresponding to a bias voltage of twice threshold). It is seen here that for small values of the constant of (3.19a) that

$$nd_{crit} \cong 2 \times 10^{11} \left. \frac{\varepsilon_{II}}{\varepsilon_I (1 - v_d^2/c_{II}^2)^{\frac{1}{2}}} \right/ cm^2$$

which is close to the value of equation (3.19). For increased loading, the proportionality with the loading factor is abandoned. Firstly, a maximum $nd_{crit}$ of $3.1 \times 10^{11}$/cm² occurs and a slight decrease in $nd_{crit}$ for increasing loading towards a value of $nd_{crit} = 2.7 \times 10^{11}$/cm².

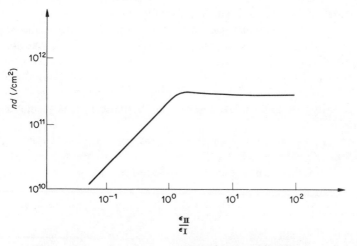

**Figure 3.1** *nd* product computed by searching for an absolute instability of the complex dispersion equation.

Although some experimental results[30] based on domain suppression with glycerine ($\varepsilon_{II} = 41\varepsilon_0$) and de-ionized water ($\varepsilon_{II} = 81\varepsilon_0$), seem to verify these computational results, the situation has not yet been entirely clarified, particularly for cases where $v_d$ becomes equal to $c_{II}$, when strong wave coupling phenomena occur. A further analysis which includes diffusion and a finite length $l$ shows, however, that the inequality (3.19) can also be valid for heavy loading.[9a]

## 2.2  DOMAIN INHIBITION BY SURFACE LOADING

The inequality of (3.19) has a whole range of implications. The first case is that of a thin active $n$-type layer of thickness $d$, as is commonly found for planar Gunn-effect structures. The $n$-layer might be fully imbedded on both sides with semi-insulating GaAs when $\varepsilon_I = \varepsilon_{II}$, or it might be deposited on a GaAs substrate with one side exposed to air, when $\varepsilon_{II} \cong (\frac{1}{10})\varepsilon_I$ as the relative permittivity of GaAs is $\varepsilon_r \cong 10$. The inequality (3.19) will then become

$$nd > 1.6 \times 10^{11}/cm^2 \qquad (3.20)$$

This means that domains will not form in such thin films which do not satisfy this condition. The condition of equation (3.20) seems to be independent of $l$, as long as $\beta\alpha \ll 1$. In order to illustrate this point, let us take the following example. A diode with $d = 1$ μm, $n = 10^{15}/cm^3$, $\omega_{cz} = 3 \times 10^{11}/s$ and $l = 100$ μm has an $nl$ product of $10^{13}/cm^2$, whereas $nd = 10^{11}/cm^2$, which is rather low for strong domain formation to occur. The condition $\beta\alpha \ll 1$ is satisfied here since $\beta\alpha \cong 0.025$ from equation (3.17). The expression (3.19) therefore represents an important limitation for planar structures, which can usefully be made to have an $nd$ of $5 \times 10^{11}/cm^2$ for domains to grow to a mature size. From a heat-sinking point of view it is, on the other hand, essential to keep $nd$ as low as possible, because $nd$ is directly proportional to the waste power produced per unit area of active layer. High power levels would be responsible for a substantial increase in layer temperature with a resulting decrease in peak-to-valley ratio (see Figure 2.1, page 6) so that the output signal quality deteriorates again.

The second case of interest is surface loading with dielectric or magnetic materials. So far, only dielectric loading has been reported for domain inhibition. In this case, expression (3.19) reduces to

$$nd > 1.6 \times 10^{11} \frac{\varepsilon_{II}}{\varepsilon_I}$$

One can consider this result also from a different point of view. Taking the large-signal time constant for domain-charge growth as given by equation (2.65), page 30 i.e.

$$T_g = R_0 C_d$$

the effective domain capacitance $C_d$ is now enlarged by a parallel capacitance which is given by the dielectric material. The penetration of the domain fields into the dielectric is taken to be approximately equal to the domain width $d_d$ so that the parallel capacitance $C_L$ due to the loading material is given by

$$C_L = \frac{\varepsilon_{II} d_d}{d_d} a_1$$

where $a_1$ is the width of the loading layer. The ratio of the time constants with and without dielectric loading is then

$$\frac{T_L}{T_g} = 1 + \frac{\varepsilon_{II} d_d}{\varepsilon_I d}$$

The suppression of domains is expected to occur when $T_g$ becomes larger than the domain transit time $t_t$. For unloaded surfaces the condition of expression (2.9), page 10, can be rewritten as follows:

$$nl = nv_D t_t > 10^{11}/\text{cm}^2$$

with $t_t = 2T_g$.

The condition for dielectrically-loaded surfaces can be expressed as follows for $\varepsilon_{II} d_d > \varepsilon_I d$:

$$nl = 2nv_D T_L > 10^{11} \frac{T_L}{T_g} \simeq 10^{11} \frac{\varepsilon_{II}}{\varepsilon_I} \frac{d_d}{d}$$

$$nd \geqslant 10^{11} \frac{\varepsilon_{II}}{\varepsilon_I} \cdot \frac{d_d}{l}$$

As $d_d/l$ can be at most equal to one, this inequality is the same as expression (3.19), which was derived more rigorously. The approximative considerations might be of doubtful value when $\varepsilon_{II} \gg \varepsilon_I$ as is suggested by the rigorous computation in connection with Figure 3.1. (See also [9a].)

One can treat the effect of surface loading as if the domain space-charge waves have been short-circuited down to a given depth. The $n$-layer is therefore rendered inactive near the surface down to a depth given by the critical $nd$ product.

The third case is loading structures where various layers of dielectric, magnetic and highly conductive layers are stacked. For example, if a thin dielectric layer is covered by a metal layer, the parallel short-circuiting capacitance due to the loading structure could even be increased, as the effective electrode distance for the stray capacitance is basically reduced. This shows that domain control can be envisaged also by conductive layers applied as loading. In fact, a suitable analysis of the small-signal dispersion equation shows that the direct application of a conductive layer to the $n$-type surface acts as domain inhibition in the same manner as a dielectric or magnetic load. This result has, of course, a less significant effect on device developments, as a conductive layer would short-circuit the electric d.c. field, which has to be established in the $n$-layer for the transferred-electron effect to occur.

## 3.3  SIGNAL-EXTRACTION BY SURFACE LOADING AND NEW DEVICE PROPOSALS

One has to consider the possibility of extremely strong, non-dissipative loading when $c_{II}$ becomes equal to $v_0$. Then the extracted domain signals are no longer attenuated, but can be transferred to a transmission line for efficient coupling. This could possibly lead to new device developments. It should be pointed out that this type of loading is not in contradiction to the results of Figure 3.1. In fact if the signals coupled out by the loading sandwich are not transferred to a transmission line, but are fed back to an inactive GaAs layer, the instability can be aided again. On the other hand, the results of Figure 3.1 do not consider the case where $v_d$ approaches $c_{II}$, because then the boundaries of the loading materials will certainly affect the conditions for the occurrence of instabilities. This whole field surely merits further work. This section outlines some possibilities of device developments. It has to be kept in mind, however, that the ideas presented here are on the whole rather speculative, firstly because a rigorous analysis has not yet been performed, and secondly because it is still difficult to find sufficiently good layers of magnetic materials, although progress is being made.

Some first device proposals based on surface pick-up were made some time ago in which ohmic or capacitive probes detected small signals when a domain was passing underneath them.[50] This technique was useful for the study of domain dynamics; however, its utilization for new device applications was impossible as the picked-up signals were too weak for a subsequent domain formation in succeeding Gunn-effect devices. Although several interesting device proposals have been advanced,[50,55] they will only be fully realized when a more efficient detection mechanism of travelling domains is available. It means that a system has to be developed which will be able to drain the microwave signals constituting the domain pulse out of the $n$-GaAs crystal. The surface probe will have to match the space-charge waves of the domains to a transmission line leading to further Gunn-effect elements.

It is possible to envisage dielectric and magnetic materials, possibly in sandwich construction, as such efficient signal-extraction probes. The inequality of (3.19) can then be used to explain the following concept.

If the actual semiconductor thickness $d$ is much larger than the value given by the $(nd)_m$ product of (3.19), then $(nd)_m/n$ represents the depth down to which the surface material couples the domain waves out of the GaAs. If $c_{II}$ approaches $v_d$, $d_m$ becomes very large. This indicates that a considerable amount of domain-signal power is coupled out of the Gunn-effect diode.

The development of single-crystal ferrimagnetic garnets has yielded magnetic materials of high saturation magnetization $4\pi M_0$ and very low spin-resonance line width $\Delta H$. If a magnetic d.c. field $H_0$ is applied in the z-direction, the positive circularly-polarized wave travelling along z shows a very high effective relative permeability $\mu_r = \mu' + j\mu''$ near the spin-resonance angular frequency $\omega_0 = \gamma H_0$, where $\gamma$ is the gyromagnetic ratio. In fact, the attenuation constant $\alpha_F$ and the phase constant $\beta_F$ are given by the following relations (see page 300 of reference 44):

$$\alpha_F = \frac{\omega}{c_0}\left(\frac{\varepsilon_r}{2}\right)^{\frac{1}{2}}[(\mu'^2 + \mu''^2)^{\frac{1}{2}} - \mu']^{\frac{1}{2}} \tag{3.21}$$

$$\beta_F = \frac{\omega}{c_0}\left(\frac{\varepsilon_r}{2}\right)^{\frac{1}{2}}[(\mu'^2 + \mu''^2)^{\frac{1}{2}} + \mu']^{\frac{1}{2}} \tag{3.22}$$

where $c_0$ is the velocity of light in free space.

The real and imaginary parts of the scalar effective relative permeability are given as follows:

$$\mu' = 1 + \frac{\omega_m T_\gamma \Delta\omega\, T}{\Delta\omega^2 T_\gamma^2 + 1} \tag{3.23}$$

$$\mu'' = \frac{\omega_m T_\gamma}{\Delta\omega^2 T_\gamma^2 + 1} \tag{3.24}$$

with $\omega_m = \gamma 4\pi M_0$, $T_\gamma = 2/\gamma\,\Delta H$ and $\Delta\omega = \omega_0 - \omega$.

Single-crystal YIG has shown experimentally an $\omega_m T_\gamma$ product of $8 \times 10^3$ (see p. 703 of reference 44). A maximum value of $\beta$ occurs for $\Delta\omega\,T_\gamma = 0.55$. With these values it is possible to reduce the wave velocity $\omega/\beta$ to values close to the Gunn-effect domain velocity $v_D$, particularly as $\varepsilon_r$ is about 16 for single crystal garnets.

If a small difference in velocities exists, wave matching can still be achieved if the circularly polarized wave travels at an angle $\theta$ away from the interface. Its wave vector has then a component in the z-direction which is equal to $\beta_s = \omega/v_d$. The applied magnetic field $H_0$ must have an angle $\theta$ with respect to the interface.

A more important improvement will, however, be achieved if the total effective permittivity can be increased. This is possible if the garnet is covered by a dielectric slab of, say, $BaTiO_3$. Such a dielectric cover 'pulls' the waves into the garnet and decreases the wave velocity.

If the circularly polarized wave has a frequency very close to $\omega_0$, (i.e. $\Delta\omega T_\gamma \leqslant 1$), the attenuation will be rather large. Such a structure can be employed for applications involving domain inhibition if the whole n-GaAs surface is loaded. By altering the applied magnetic field $H_0$, such a device can be switched easily between the two states of domain formation and domain inhibition, firstly because equation (3.19) is very sensitive to small changes of $c_{II}$, if $c_{II}$ is near $v_d$, and secondly because $\beta_F$ of equation (3.22) changes very abruptly for small variations in $\omega_0$ if $\Delta\omega < 1/T_y$. This can possibly be utilized, for example, for logic gates in connection with pulse communication techniques. An advantage of the devices would be their insensitivity to reflected signals (see sections 2.4 and 4.3), as spurious domains cannot be nucleated by reflected pulses because these do not produce a change in applied magnetic field $H_0$, i.e. feedback is avoided because the input signals are applied as current pulses to the magnetizing coils.

On the other hand, if only part of the surface is garnet-loaded, the output pulses produced by domains can be shaped continuously by a variation in $H_0$. In fact, it should be possible to nucleate a new domain near the cathode if the first domain is drastically reduced in size near the surface load, and interesting new functional devices can be envisaged.

If $\Delta\omega > 1/T_y$, $\alpha_F$ is reduced so that one can use garnet loading for coupling out domain signals in order to employ them for further signal processing. The surface load would therefore act in the same way as capacitive or ohmic probes, except that the coupling is much more efficient as matched conditions can be set up. In fact, good matching could lead to complete domain extinction. As $\beta_F$ can be decreased by increasing $H_0$ (see equation (3.22)), suitable shaping of the garnet can be used in order continuously to increase $\beta_F$ away from the interface so that the wave is finally matched to the entrance of a transmission line (e.g. a microstrip line), connected to the end of the garnet. In this way maximum domain-signal power can be obtained which will be sufficiently large for domain nucleation in subsequent Gunn-effect pulse devices.

Using the very approximative analysis of dielectric surface-loading as outlined above, one can extend the treatment also for magnetic surface-loading. The magnetic material with the effective permeability relevant for the wave which can couple to the space-charge wave (namely the right circularly-polarized wave), is represented by an equivalent inductance $L$ in parallel with the domain capacitance. The growth time of the domain is then increased and an expression can be obtained which represents a rough approximation of equation (3.19). The simplified model can be used in order to study the effect of a resonance which can be set up by domain capacitance and magnetic-material inductance. The resonance angular frequency is

$$\omega_F = \left[\frac{1}{L_L C_d} - \frac{1}{4C_d^2 R_0^2}\right]^{\frac{1}{2}}$$

where $R_0$ is the low-field diode resistance,

$\qquad C_d$ is the domain capacitance,

$\qquad L_L$ is the equivalent loading inductance $= w\mu'$,

$\qquad w$ is the transverse width of the interface.

If $\omega_F$ is equal to the fundamental domain frequency, the signal, decoupled from the domain, will be applied to the diode again after a phase delay of about 180°. This structure might be suitable for Gunn-effect memory applications (see section 4.2). This has been considered by employing a Gunn-effect diode in a microwave resonator, biased with subthreshold voltage for domain nucleation. A small input signal raised the diode field above nucleation threshold and a domain was formed. As the voltage swing of the resonator produces a further domain after the previous one has reached the anode, the oscillation is maintained although the bias voltage is slightly below threshold, and the signal is stored. It can only be erased when a negative pulse is applied to the diode. Gunn-effect diodes with magnetic surface-loading can possibly be envisaged for such memory applications. The advantage is given by the stronger and more direct coupling between load and diode. This should produce a larger bias hysteresis than the cavity memory, i.e. the difference between nucleation threshold voltage and erase voltage is increased, giving more reliable operation.

In the case of resonance loading the application of a dielectric on top of the magnetic material is again of advantage as this effect will 'pull' more wave field into the garnet and this will increase $L_L$.

In connection with wave coupling at the surface of the Gunn-effect semiconductor, spin waves should also be of interest. As the magnetic dipoles of a magnetic material tend to align each other, wave motion is possible. The dipoles precess about the externally-applied magnetic field vector $\bar{H}_0$, and the magnetic field of an electromagnetic wave produces small disturbances of the dipole alignment. The 'disturbing' magnetic field is produced by the space-charge waves of the Gunn-effect domain near the interface. If the phase velocity of spin waves is comparable to that of a domain, coupling can be very efficient. The dispersion diagram of spin waves is shown in Figure 3.2. Their 'surface waves' occur only with thin rods. The 'volume waves' have, on the other hand, wave numbers which can be larger than the space-charge waves of $n$-GaAs, whose domain velocity $v_D$ has been entered in Figure 3.2 by a dash-dotted line. The cross-points of $v_D$ with the dispersion curves of spin waves give the conditions for efficient matching. It can be seen that coupling should be possible with spin waves travelling both parallel and perpendicular to the applied magnetic field $H_0$. As the spin waves under consideration here have an r.f. magnetic-field component perpendicular to the d.c. magnetic field, and a longitudinal component along the direction of propagation (see p. 318–320 of reference 44) it is possible to have a range of coupling systems. In fact, waves can be coupled off which have a wave vector either (a) in direction of domain velocity, or (b) perpendicular to domain velocity and parallel to the interface, or (c) perpendicular to the interface, by arranging the direction of $\bar{H}_0$ suitably. Case (b) should

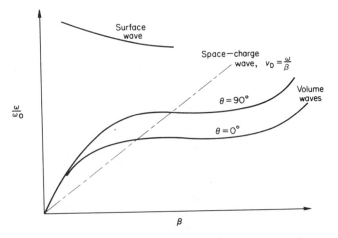

**Figure 3.2**   Dispersion diagram of spin waves ($\theta$ is the angle between the wave vector and $\bar{H}_0$).

be of great interest for the development of Gunn-effect neuristor logic,[55] where various parallel and antiparallel Gunn diodes are surface-coupled to each other. These diodes can then be built in coplanar technology, which is very suitable for mass-production techniques.

A range of device applications has been outlined here which require the use of relatively advanced magnetic materials for surface-loading of Gunn-effect domains, and considerable further work is required in order to produce experimental verification of the theoretical proposals.

## 3.4   CAPACITIVE AND OHMIC SURFACE PROBES

When the area of loaded surface is reduced to a very small value compared with the device's total free surface, signal power can be extracted for analysing purposes. This power is, of course, very weak because the probe size is small. However, it permits one to study the domain parameters at one position. By moving the probe along the device surface, such details as domain growth and speed can be obtained.

It is, of course, not then possible to use the signals obtained to nucleate domains in further devices, as the signal power is too weak, i.e. it is below the input level required for Gunn-effect regenerators. As it is of great value for analytical purposes, and could possibly be used for displaying logic output signals directly, it is described here in further detail.

Both capacitive and ohmic probes have been employed; however, an ohmic probe has to be alloyed into the surface and is therefore not movable. A capacitive probe has typically the shape as demonstrated in Figure 3.3. In this example, a very thin (typically 3 μm) aluminium

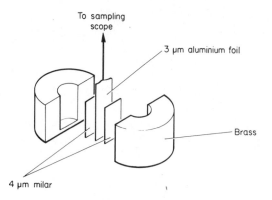

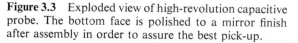

**Figure 3.3** Exploded view of high-revolution capacitive probe. The bottom face is polished to a mirror finish after assembly in order to assure the best pick-up.

foil is clamped between two semicircular brass ground plates with two 4 μm milar sheets for insulation. The brass plates have the shape shown in Figure 3.3. If the cylindrical cut-away parts were not provided, the high shunt-capacitance of the line would short-circuit the picked-up signals. The shunt capacitance should be around 1–2 pF, giving, in conjunction with a 50 Ω load, a rise time of about 100 ps. The end face of the probe assembly has to be brought as close to the diode surface as possible, as the fringing fields of the domain instability penetrate only a very small distance into the free space, typically the same amount as a wavelength of the space charge instability, i.e. the domain width. It is therefore necessary to lap and polish both the end of the probe after assembling the sandwich, and also the surface of the GaAs device (after it has been embedded into a suitable material, for example Araldite). In order to ensure constant separation between probe and GaAs surface, some very thin insulating film or sheet can be used.

The aluminium foil can be connected either directly to a high-impedance probe of a sampling oscilloscope in order to measure the probe voltage, or to a 50 Ω, low-impedance sampling unit in order to obtain the time derivative of the probe voltage.

Device developments based on small surface probes can be foreseen when using ohmic contacts in order to influence the current distribution inside the GaAs crystal. Some first experimental verification with relatively long samples was achieved already in 1967, but future developments for gigapulse rates have to rely on photolithography with planar structures. The current distribution inside the device can be affected by connecting the surface contacts to a resistive shunt path. When the domain travels underneath the shunt path, it experiences a decreased electric field there so that the excess domain voltage $V_D$ is reduced and the domain current

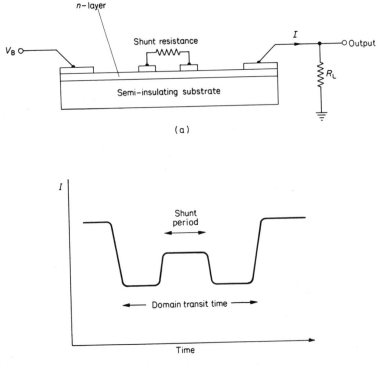

**Figure 3.4** External shunt resistor generates wave functions: (a) circuit, (b) current wave for shunt resistor in series to Gunn-effect domain.

pulse across the series load resistor $R_L$ is reduced in amplitude (see Figure 3.4). If now the resistor is connected in series with a current switch, the resulting domain-current wave is an indication of the state of the switch, i.e. when it is open no shunt period is visible. The domain can now be used to read out the state of external switches. For example, binary pulses can be inserted into the output waveform by simply opening or closing the resistive paths between contacts, and applications such as time division multiplexing can be envisaged.

External resistors made of photoconductive material change their value in proportion to the intensity of illuminating light. This possibility could provide a very high-speed electrical read-out of a complex digital optical pattern or of a continuous image. Similarly, the voltage signal applied between two contacts when a domain is travelling between them can be employed to activate an array or a continuous layer of electronic components, such as semiconductor switches or light-emitting elements. The actuated devices can be directly attached to the semiconductor devices, for example in the form of light-emitting heterojunctions grown epitaxially directly onto the free surface of the active GaAs, or they can be connected to it via conducting leads.

## 3.5 MATERIAL STABILITY ON SURFACES

Impurities not only migrate across single crystal semiconductors due to diffusive effects, but can also be pulled across relatively long distances by strong electric fields if the impurity atoms are ionized. For example, tin has been found to migrate across bulk GaAs because it becomes an ionized donor which is transferred from a positive anode to a negative cathode. It has been found that the high fields associated with Gunn-effect domains are often responsible for such material migration. An estimate of the speed of migration for tin has given a transfer distance of several Ångströms per domain transit.[17] It is therefore important not to employ near device surfaces any materials which exhibit such a high-field migration phenomenon.

In fact, this detrimental effect of high fields was strongly enhanced when a high-field component was directed along the GaAs surface. Material then migrated along the surface, presumably because the bonding forces of the single crystal are altered there with the result that impurity atoms can travel more easily. Once material had in fact been pulled across from the anode to the cathode, the device was usually short circuited because a metal channel was formed along the active layer. It is therefore essential to select alloying materials which are stable regarding bulk migration. Many of the metals suitable in connection with this consideration still unfortunately show surface migration, and this has been found particularly so with the very popular alloy In–Ge–Ag (see section 4.4). It is thus imperative to avoid large field components along the surfaces near the metal anodes.

The exact nature of metal migration along surfaces has not yet been established. There is a suggestion that the initiation of migration is caused by local lattice-heating due to recombination of injected holes. This theory has been supported by experimental evidence of recombination radiation near the anode electrode.[63] Such material migration has, however, also been found along the surfaces of semi-insulating material, as shown in Figure 3.5.[10] Here, the planar GaAs device had an 11.6 μm active-layer thickness on a semi-insulating substrate. The carrier density in the epitaxial *n*-layer was $10^{15}/cm^3$, the mobility 7000 $cm^2/V$ s, and the orientation of the surface (100). The contacts were produced by evaporating In–Ge–Ag (In + Ge 11 per cent, Ag 89 per cent); the interelectrode distance was 100 μm. After alloying of the contacts, the structure was selectively etched 14 μm deep with $H_2SO_4 + H_2O_2 + H_2O$ (1:1:3) so that a mesa device resulted, the active layer of which is much

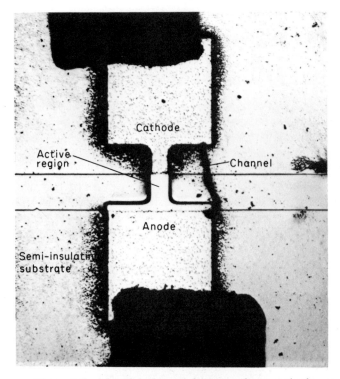

**Figure 3.5**   Material channel from anode to cathode along the semi-insulating surface of the substrate.

larger at the anode than at the cathode in order to avoid a high surface field near the anode.

Metal migration was found to occur not only along the semiconducting surfaces, where high electric fields due to Gunn-effect domains can occur, but also along the semi-insulating surface (see Figures 3.5 and 3.6(a)) where the resistivity is $10^8$ $\Omega$ cm. It was found that a sharp electrode edge initiated migration due to the high field existing there. The resulting black line developed within seconds of applying a pulsed bias voltage of around 150 V (pulse duration 40 ns, pulse repetition 100 Hz). This channel crossed the interelectrode distance along the uncovered semi-insulating surface, as can be seen in Figure 3.5. As soon as the line had connected both electrodes, no domain oscillations occurred any more.

A microprobe analysis on this diode shows the channel contains primarily Ag (Figure 3.6(b)), but no indication of In or Ge could be found. One has to take into account, of course, that the InGe content is much smaller than Ag in the contact electrodes. In fact Ag was found to be removed from the anode corner where material migration started.

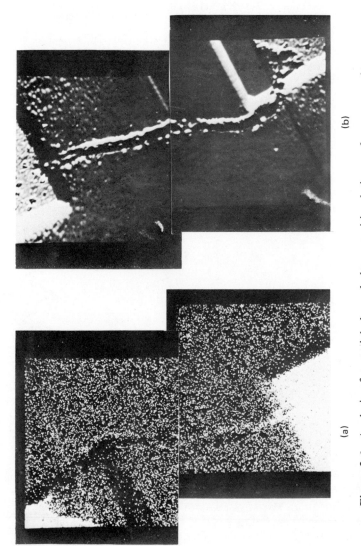

(b)

(a)

**Figure 3.6** Analysis of material channel along semi-insulating surface; upper electrode: cathode; lower electrode: anode; magnification: ×1200; (a) micrograph by scanning electron microscope (b) micrograph by microprobe analyser showing Ag content by brighter parts. As can be seen, Ag is missing near the anode and is present in the channel.

The resistance of diodes with channels decreases for some time immediately after the formation of a channel, obviously due to anode material still being transported, until some equilibrium is established after a few minutes.

Many of these problems can, of course, be avoided by using a regrown $n^+$ contact as produced by liquid epitaxy.

# 4: Logic Devices Based on the Gunn-effect Domain

## 4.1 PULSE-SIGNAL REGENERATION

By biasing Gunn-effect domain devices in resistive circuitry just below the transferred-electron threshold, a small input signal of the right polarity applied to the device can raise the field in the GaAs layer sufficiently to nucleate a domain. As shown in Chapter 2, once this instability has been nucleated it will continue to travel towards the positive electrode as long as certain device parameters are given and the bias voltage is above the domain extinction value. The domain instability will disappear only when it reaches the anode electrode. During domain transit the current is reduced, maximally, by up to 60 per cent, and the resulting pulse-signal voltage produced across a load resistance can be substantial. As the input signal can be relatively small, such a device shows pulse-regeneration gain.

The maximum pulse gain can be found by estimating the field increase $\Delta E$ required to switch the diode from the subdomain state to reliable domain generation. Experimental results[35] have shown a power gain of up to 24 db for a 50 $\Omega$ load resistor, a diode low-field resistance of around 200 $\Omega$, $l$ around 600 $\mu$m, and with pulsed bias voltages so that the active layer remained at room temperature. When planar devices were supplied with suitable heat-sink structures and operated with d.c. bias voltages, the active-layer temperature was, of course, rather elevated so that the pulse-signal gain was greatly reduced. Experimental structures have then shown 6 db power gain for two-terminal devices[14] and more than 8 db for Schottky-gate trigger devices,[58] see section 4.3.

The theoretical value of $\Delta E_i$ can be estimated by taking the noise fluctuations of the carrier density. Here an important noise phenomenon could be shot noise, although other noise effects might also be important. Shot noise produces field fluctuations across the nucleating region. The resulting expression shows that these fluctuations are proportional to $n^{\frac{3}{4}}$, so that the gain decreases with increasing conductivity.[21] Further experimental efforts have to clarify these phenomena considerably.

## 4.2 SIMPLE TWO-ELECTRODE LOGIC COMPONENTS

As the Gunn-effect domain exhibits a threshold effect, it can easily be used to set up logic gates. Firstly, AND gates have been made by connecting the input terminal of a Gunn-effect diode regenerator to two separate input terminals via series resistances $R_i$ (Figure 4.1). A signal from one

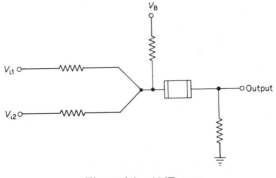

**Figure 4.1**   AND gate.

input terminal alone does not suffice to raise the field above $E_t$, as the signal is reduced by $R_i$. Only the application of two signals simultaneously to each terminal succeeds in nucleating a domain.

The removal of $R_i$ or a slight increase in bias voltage enables one to produce an inclusive-OR gate.

Two Gunn-effect diodes operating in parallel onto one series resistance $R_1$ can be made to act as an exclusive OR (Figure 4.2). This is because the simultaneous occurrence of both input signals (applied each to one Gunn-effect diode) causes the voltage in $R_1$ to increase to such a value that the diode field remains below $E_t$. The same effect is achieved if two small-sized anode electrodes are placed opposite a long cathode[21] (Figure 4.3).

Inverters can be produced by having the bias voltage to the above inclusive OR circuit slightly above the threshold value. Unless an input signal of opposite polarity to the bias voltage is applied, a domain is

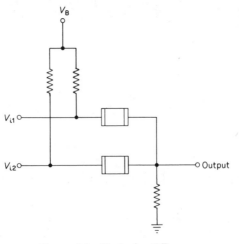

**Figure 4.2**   Exclusive OR.

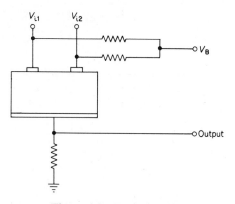

**Figure 4.3** Exclusive OR.

formed. In order to have domains occurring only when a signal is to be expected (but not arriving), the bias supply could be of subthreshold d.c. level plus a voltage wave in pulsed form at the signal repetition frequency and phase, as shown by Figure 4.4. An input signal of opposite polarity then neutralizes the bias trigger pulses.

A Gunn-effect memory can be arranged either by placing two series-type regenerators in a closed loop or by connecting a short section of open-ended delay line to the output of a single regenerator.

With the help of the components described logic systems can be set up, and a first approach was a feasibility study regarding a shift register as described in the next chapter.

Instead of the two-level system described so far (where the 'on' condition is represented by a domain pulse and the 'off' condition by the absence of a pulse) there are advantages if a three-level logic[22] can be used. Here, say, a negative pulse represents 'off', a positive one represents 'on', and the absence of a pulse shows that no information is flowing. The attraction is that the 'off' signal is different from the state of 'no information flow', a feature which is often useful for the operation of monostable logic such as that described here.

The basic unit is given by a series combination of two Gunn diodes as shown in Figure 4.5. The bias voltages are applied in such a way that the middle point of the chain is at earth when no domain is present. They are just below threshold for domain nucleation so that a small input signal applied to the middle point will nucleate a domain in one of the diodes. If the signal voltage is positive, the diode with the negative bias voltage nucleates a domain and a positive regenerated output pulse is produced. On the other hand, a negative input signal triggers the other diode and a negative output pulse results.

In order to establish that the usual logic gates can be set up by such three-level units, suitable diode pairs were connected on one strip-line

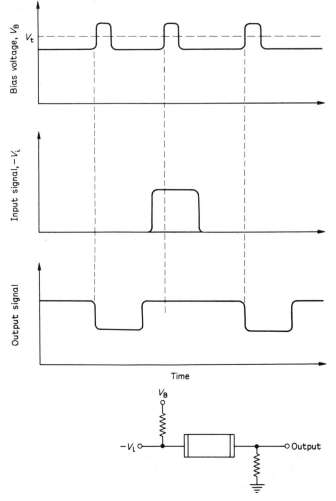

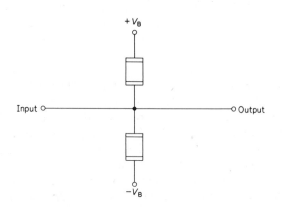

**Figure 4.4**  Inverter (note that the input signal has a negative polarity).

**Figure 4.5**  Basic unit of three-level Gunn-effect logic.

unit. A voltage gain of more than 20 db has easily been obtained for a diode peak-to-valley ratio of about 50 per cent when the devices were employed under pulsed bias voltages. D.C. biasing would, of course, reduce the gain value again owing to the higher operating temperatures.

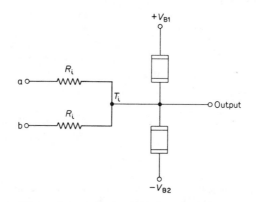

**Figure 4.6**  AND gate of three-level logic.

An AND gate can be produced by employing two input resistors $R_i$ as follows (see Figure 4.6). The signal amplitude $V_s$ has to be larger for one polarity (representing the 'off' condition) than for the other one. To give a numerical example, $V_s$ has to be either $+5$ V (for 'on') or $-8$ V (for 'off'). This can easily be arranged by selecting diode pairs, where one diode with a high low-field resistance together with a parallel resistance is paired with a diode of small low-field resistance. The effective peak-to-valley ratio will then be smaller for the first diode than for the latter one. If the resulting signals are applied to terminals a and b of the AND gate of Figure 4.6, the operation of Table 2 can be obtained. The threshold for negative pulses has to be about $-1.0$ V and can be as large as about $+4.5$ V for positive ones for the numerical example given. It is seen that the circuit acts, in fact, as an AND gate.

TABLE 2
Operation of AND Gate, Shown by Numerical Example

| Case | Signal pulse voltage (V) at a | b | Resulting signal at $T_i$ (V) | Domain nucleation in diodes | Output pulse polarity | Type of binary signal |
|------|------|------|------|------|------|------|
| 1 | $+5$ | $+5$ | $+5$ | B | $+$ | on |
| 2 | $+5$ | $-8$ | $-1.5$ | A | $-$ | off |
| 3 | $-8$ | $+5$ | $-1.5$ | A | $-$ | off |
| 4 | $-8$ | $-8$ | $-8.0$ | A | $-$ | off |

If we reverse the binary significance of the signal pulses, we can employ the circuit of Figure 4.6 also as an OR gate. We have to reverse the polarity of both the input and output pulses by the circuit of Figure 4.7, which acts as an 'inverter'.

This inverter circuit is basically a parallel connection of two two-level series regenerators. An input signal of a given polarity triggers a domain only in one of the diodes. The resulting output pulse of opposite polarity is supplied to the output terminal via relatively high resistances $R_c$. As an individual diode gives relatively high values of regenerator gain, $R_c$ can be made quite large, and one still obtains sufficiently strong output signals for domain nucleation in subsequent Gunn-effect diodes. If $R_c$ is sufficiently high, the circuit of Figure 4.7 can be redrawn to give Figure 4.8 where the output is the arithmetic sum of the potentials at a and b. One can see that for very large values of $R_2$, approximately the same current is applied to both resistors $R_1$. A domain pulse in one of the diodes therefore produces pulses at a and b of the same amplitude and of opposite polarity, so that the resulting output signal is zero. This argument suggests that $R_2$ must be as small as possible.

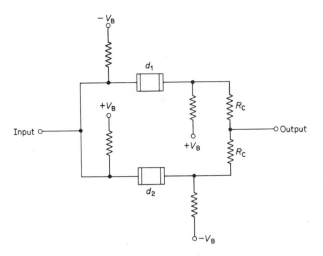

**Figure 4.7** Three-level inverters $d_1$, $d_2$ are Gunn-effect diodes, $V_B$ is a bias voltage.

Assuming a peak-to-valley ratio of 50 per cent for each diode, the resulting signal amplitude is given by the following expression, where $V_t$ is the threshold voltage for domain nucleation for each diode, and $k$ is the

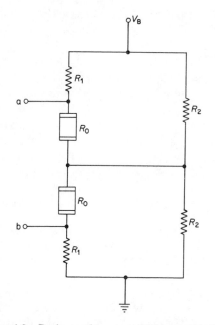

**Figure 4.8** Redrawn inverter of Figure 4.7. $R_0$ is the low-field resistance of the Gunn-effect diode, and a and b are terminals giving the output signals.

coefficient indicating the signal reduction due to the resistors $R_c$.

$$\frac{V_s}{V_t} = k\left[\left(\frac{R_1}{R_0} + 1\right)\frac{1}{R_2/R_1 + 2R_0/R_1 + 2}\right.$$

$$\left. - \frac{1}{4}\left(\frac{R_1}{R_0} - \frac{R_1/R_0}{1 + 2R_0/R_2 + 2R_1/R_2}\right)\right] \quad (4.1)$$

The value of $V_s$ will be the largest for very high resistances $R_1$. For analytical convenience we assume here that $R_1 = R_0$ and find $V_s/V_t = k/(4 + R_2/R_1)$, which indicates that in fact very small values of $R_2$ give largest $V_s$. Of course, a compromise has to be found because small $R_2$ results in increased power dissipation, which is undesirable. If $R_c$ is taken to be more than five times larger than $R_0$ and $R_1$, its effect on the operation of the circuit is negligible. Assuming as reasonable values $R_0 = R_1 = 50 \, \Omega$, $R_c$ can then be 300 $\Omega$. A subsequent pulse regenerator of the type of Figure 4.5 has to be triggered by the signals originating from the $R_c$ combination. If the low-field resistance $R_{0r}$ of the regenerator diodes is taken to be 50 $\Omega$, a signal amplitude of $\frac{1}{24}V_t$ results, which is well above the minimum signal value required for successful Gunn-effect regenerator operation as given by the power gain of 24 db presented in section 4.1. The circuit can be further improved by selecting regenerator diodes of high $R_{0r}$ inserted in a correspondingly high impedance line, as $R_c$ can then be reduced.

Further work still is required before all possibilities are fully exploited; however, the results obtained so far are encouraging.

The two-electrode Gunn-effect pulse regenerator and its logic elements have the unfortunate disadvantage that they show no directionality. A reflected pulse arriving at the regenerator output can nucleate a domain in the same way as a true input pulse. This feature represents a serious handicap for pulse-processing systems, and it is useful to develop reflection-insensitive Gunn-effect regenerators, as described in the next section.

## 4.3　REFLECTION-INSENSITIVE REGENERATORS

These cannot be set up by simply employing two-terminal elements. They have to be designed by incorporating at least three electrodes, in the same way as a transistor represents a three-terminal device. There is firstly a series-type three-terminal regenerator,[23] this results in (a) a rather long diode structure giving a reduced pulse-processing speed, and (b) a contact geometry which is difficult to evaluate correctly. The recently-observed fast spread of Gunn-effect domains in directions perpendicular to the domain trajectory[40,62] makes a parallel-type arrangement of a three-terminal regenerator[24] possible. This component has a cathode electrode c opposite two anodes a, where all three electrodes are separated from each other by active semiconducting GaAs. An an example, a possible realization is shown by Figure 4.9 which represents a planar

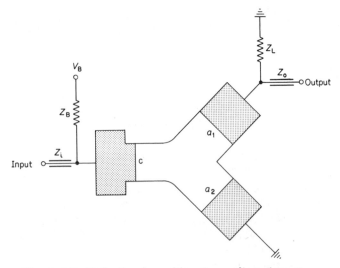

**Figure 4.9**　Reflection-insensitive Gunn-effect element (c—cathode, a—anode).

Gunn-effect structure of, say, a 10 μm $n$-layer on a semi-insulating substrate, structured by photoresist technology. The input signal is applied via a line of characteristic impedance $Z_i$ to the cathode, which is also biased with $V_B$ via a bias line of impedance $Z_B$, where $Z_i \neq Z_B$. One anode, $a_1$, is connected to earth directly, the other one, $a_2$, is connected to earth via a load impedance $Z_L$. The output signal is taken off from $Z_L$. The values of $V_B$ and $R_L$ have to be such that the field between $a_1$ and c is just below the threshold for domain nucleation, $E_t$, and the field between $a_2$ and c is much smaller than $E_t$ but well above the domain-extinction field $E_e$. Figure 4.9 also shows how the other requirements

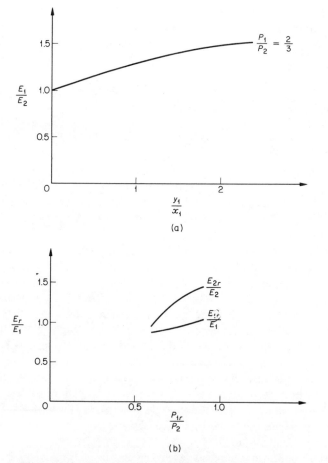

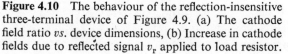

**Figure 4.10** The behaviour of the reflection-insensitive three-terminal device of Figure 4.9. (a) The cathode field ratio *vs.* device dimensions, (b) Increase in cathode fields due to reflected signal $v_r$ applied to load resistor.

of planar Gunn-effect devices can be satisfied, namely a field near c lower than near a in order to avoid migration of anode-contact material, and large metal surfaces for thermocompression bonding of lead wires.

One can show by a relatively simple analysis concerning potentials, that the field ratio $E_1/E_2$ near the cathode depends on the ratio of the distance $2y_1$ between the anodes to the average distance $x_1$ between anode and cathode as given by Figure 4.10(a). The field below $a_1$ is denoted by $E_1$ and that below $a_2$ by $E_2$; $P_1$ and $P_2$ are the potentials of the anodes $a_1$ and $a_2$ respectively. One has to find suitable values of $y_1/x_1$ and $P_1/P_2$ so that $E_1$ is just below the threshold field $E_t$, i.e. about 3000 V/cm, and $E_2$ is half-way between $E_t$ and $E_e$, the domain extinction field, e.g. 2500 V/cm. Taking $Z_L$ to be 50 $\Omega$, and the low-field resistances between c and $a_1$, and c and $a_2$ to be both 100 $\Omega$, the potential between c and $a_1$ is two-thirds the potential between c and $a_2$. This means, approximately, that $P_1/P_2 = \frac{2}{3}$. Taking $y_1/x_1 = 0.5$, the suggested value ratio can, in fact be set up, as can be seen from Figure 4.10.

Finally, one has to show that a reflected signal $v_r$ applied to $R_L$ does not bring $E_1$ or $E_2$ above $E_t$. The effect of $v_r$ is to increase $P_1$ so that a new value $P_{1r}$ occurs. For $P_{1r}$ we can calculate the corresponding terms $E_{1r}$ and $E_{2r}$ which are the new values of $E_1$ and $E_2$ respectively. Figure 4.10 shows $E_{1r}/E_1$ and $E_{2r}/E_2$ vs. $P_{1r}/P_2$.

This indicates that for $P_{1r}/P_2 = 0.9$, $E_1$ is only increased by 7 per cent and $E_2$ by 23 per cent so that both are still below $E_t = 3500$ V/cm. This value of $P_{1r}/P_2$ indicates that the reflected signal can be so large that the voltage across $Z_L$ is reduced by 66 per cent, without causing any domain nucleation.

Domains can only be created by an input signal applied to the cathode when $E_1$ is increased above $E_t$. Once a domain is nucleated, the current through the device is reduced by $I_D$. This current reduction $I_D$ is shared between $a_1$ and $a_2$, whereas only the part going through $a_1$ causes an output pulse to occur across $Z_L$. The pulse-regeneration gain will therefore be reduced by more than 50 per cent. However, as very high gain values have been observed with two-terminal devices, the resulting gain will still be attractive.

Once the domain reaches the junction between $a_1$ and $a_2$, the domain will be divided into two halves. The domain part proceeding towards $a_1$ can be used to estimate the pulse amplitude obtainable across $Z_L$.

Recently, experimental results have become available regarding such a reflection insensitive device.

An easier solution regarding reflections has been employed successfully so far by using a two-electrode Gunn-effect element together with a fast-response Schottky diode as shown in Figure 4.11. A Schottky junction consists of a semiconductor crystal supplied with a rectifying

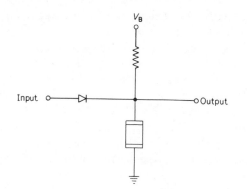

**Figure 4.11** A pulse regenerator using a unidirectional junction diode, which can be a Schottky diode in view of high speed.

metal contact. When a voltage is applied in forward direction, electrons are injected from the semiconductor into the metal where they have elevated carrier temperatures. Therefore these devices are also sometimes called hot-electron diodes. When a reverse bias is applied, a depletion layer is formed at the semiconductor-metal interface. The depletion process can occur very fast because no minority carriers have to be removed as required for *np*-junction devices. Schottky devices can therefore also operate fast, although the Schottky-junction capacitance limits the application for very high pulse rates. This combination with a two-electrode Gunn-effect element can therefore usefully be employed for the lower gigapulse rates. An integrated unit based on two Gunn-effect diodes and two Schottky diodes including the required load resistors has been reported.[45] It operates at 2 gigapulse/second.

A further improvement can be obtained by using a separate capacitive domain-trigger electrode which can conveniently be a reverse-biased Schottky electrode.[58] Such a possibility seems to improve the trigger sensitivity and thus the regeneration gain, as was found experimentally. However, the proposed property of reflection insensitivity is only small, because domains are usually nucleated at the anode side of the Schottky electrode, where a reflected signal applied to the anode would equally cause domain generation.

Regarding domain triggering by Schottky electrodes, usually deposited on planar structures between cathode and anode, one has of course two different types of mechanisms. Firstly, a positive trigger pulse (with respect to cathode potential) switches the gate into forward conduction. The additional current increases the current density near the cathode until a domain is nucleated between cathode and gate. This mechanism is equivalent to that of domain triggering by an ohmic contact, and acts in the same manner as the series-type three-terminal regenerator mentioned above. Unfortunately, this device requires an intermediate electrode with an effective low-impedance bias for reflection-insensitivity to be achieved, and the rectifying property of a Schottky gate does not permit one to obtain this in the rest condition when the gate would be rectifying.

Secondly, a negative pulse switches the Schottky gate into the reverse direction and a space-charge layer is formed which reduces the cross-section of the current-conducting channel. Thus, the field in the gate region increases until a domain is nucleated under the gate. A figure of trigger capability has been defined here as the ratio of the amplitude of the output negative current to the minimum trigger pulse voltage. This would give pulse-voltage regeneration gain when multiplied by the load resistance applied to the anode.

A similar figure can be defined for the trigger capability of a signal of correct polarity applied to the anode electrode when a reflection some-where along the output line could nucleate spurious, unwanted domains. Some simple algebra (see the appendix to this chapter) based on a suitable approximation regarding the effect of the gate field, gives the ratio of the figure of gate trigger capability to that of anode trigger capability, i.e.

$$\gamma \cong l_t \frac{m_t - 1 + k' - m_t x}{k'} \cdot \frac{m_t E_t}{2\phi_p x(1 - x)} \qquad (4.2)$$

where $E_t$ is the threshold field for domain nucleation;

$\quad k' = 1 - k$, is related to the peak-to-valley ratio;

$\quad m_t = E_g/E_t$, with $E_g$ the critical field under the gate for launching
$\qquad$ domains;

$\quad x = y_d/d$, with $y_d$ the depth of the depletion layer under the gate
$\qquad$ and $d$ the thickness of the $n$-layer;

$\quad \phi_p$ is the pinch-off voltage, which when applied to the Schottky
$\qquad$ gates makes $y_d = d$;

$\quad l_t$ is the distance between gate and anode.

Although experimental results have shown a $\gamma = 0.4$, that means the anode-trigger capability is still larger than the gate trigger capability,[56] for optimum conditions (for 1 $\Omega$ cm material with, for example, $d = 5$ $\mu$m, $l_t = 100$ $\mu$m, $m_t \cong 1.06$ and with an optimum value of the gate bias) a value of $\gamma = 3.6$ can be predicted. This would mean that a limited amount of directionality can be achieved.

The feedback signal to the input circuit due to the formation of any domain is negligible and was found to be smaller than 30 db. D.C. operated regenerators have been developed by Japanese workers.[58] The input pulse amplitude needed to be 0.2 V, whereas the output pulse amplitude was 0.55 V for a single device and 1.2 V for a double-device layer on one substrate. The sample length was 50 $\mu$m, the output pulse width 0.6 ns, the rise and fall time 0.15 ns, and the low-field resistance for a single device 750 $\Omega$. The temperature increase was estimated to be 120 °C.

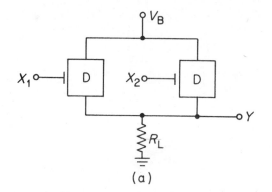

(a)

Output pulse (at Y)

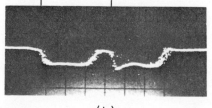

(b)

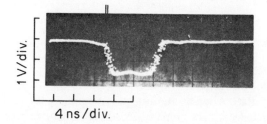

1 V/div.

4 ns/div.

**Figure 4.12** (a) Pulse processing circuit consisting of two Schottky-trigger-gate Gunn diodes D. The output is given by (b) for a delay between the input pulses, and by (c) for simultaneous application of the input signals at $x_1$ and $x_2$.

Experimental studies of numerous logic circuits have demonstrated the usefulness of this approach for the lower gigapulse rates. The circuit of Figure 4.12(a) processes input pulses applied to $X_1$ and $X_2$. Figure 4.12(b) shows the output for two input signals with a delay between them, and Figure 4.12(c) gives the case when both input signals occur simultaneously. If the output Y from this circuit is connected to a subsequent Schottky-gate regenerator which triggers only due to large signals as obtained when two input signals are applied simultaneously to the circuit, an AND gate

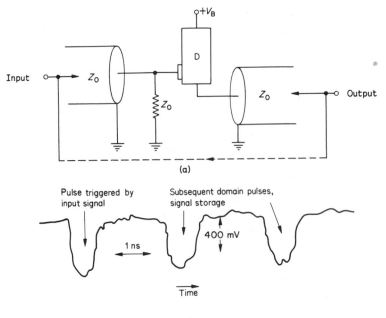

(a)

Pulse triggered by
input signal

Subsequent domain pulses,
signal storage

1 ns

400 mV

Time

(b)

**Figure 4.13** (a) memory loop consisting of a Schottky-
trigger-gate Gunn diode D. (b) Its output signal for one
input signal pulse applied.

is achieved. On the other hand, if the regenerator is triggered by any
input pulse, also by a pulse representing a single input signal to the
summation circuit, an inclusive OR is obtained.

A memory loop is given by Figure 4.13(a) where $V_B$ is again just below
the threshold value. After an input signal is applied, the resulting output
pulse is re-applied to the Schottky gate. A continuous wave as given by
Figure 4.13(b) is obtained, until an 'erase' signal cancels the effect of a
pulse on the gate and no further domains are nucleated.

A neuristor line can easily be set up by using the details of Figure 4.14.
Here a pulse signal travels without any loss in signal amplitude. This

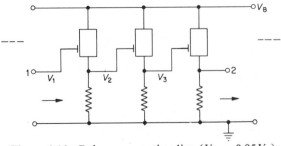

**Figure 4.14** Pulse regeneration line ($V_B \sim 0.95 V_T$).

circuit can be easily integrated. The Schottky gate is self-biased. An inhibitor or NOT gate is shown by Figure 4.15. Here the output signal O has either a positive polarity if the input signal at $S_1$ arrives first, and a negative polarity if the input signal at $S_2$ is applied first. Once a signal has been nucleated by a signal at $S_2$ or $S_1$, a subsequent signal can no longer generate any domain.

Another memory circuit is shown by Figure 4.16, together with experimental wave shapes (a)–(d). An input signal (a) starts continuous domain nucleation (b) because the capacitor $C$ maintains the cathode electrode near earth potential even after the collapse of a domain so that the device field is sufficiently high for new domain nucleation. An input signal (c) of opposite polarity can be employed to erase the stored signal (d) because further domain nucleation is then prevented.

These examples might suffice to demonstrate that numerous logic components can indeed be developed. An impressive monolithic integration of Gunn-effect logic was reported by Mause *et al.*[45a] The next task is to show that these components can be employed to form a logic system.

## 4.4  RELEVANT TECHNOLOGY

It is not intended to give an exhaustive review of Gunn-effect technology here. This is a very extensive field and would require the space of an entire book on its own. Numerous good reviews[7,52] and other papers[9,47,64] have been published recently, and the reader is referred to these. To produce GaAs single-crystal layers, ohmic contacts with as low a metal-contact resistance as possible, and contact stability under high fields—just to name a couple of important aspects—is a difficult art which

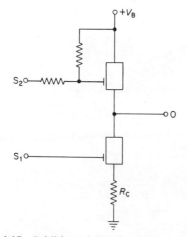

**Figure 4.15** Inhibitor (above) and (on page 86) its input and output waveforms. (Output waves for changing delays between input signals show that for a small delay the first signal arriving dominates).

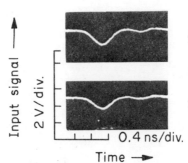

Signal applied to S₂

Signal applied to S₁

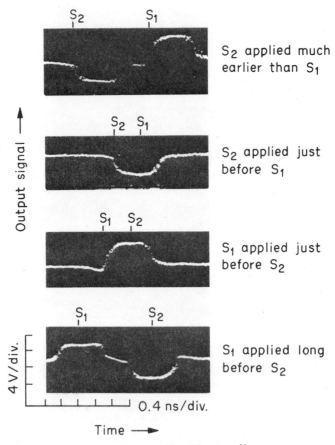

S₂ applied much earlier than S₁

S₂ applied just before S₁

S₁ applied just before S₂

S₁ applied long before S₂

**Figure 4.15** (*Continued.*)

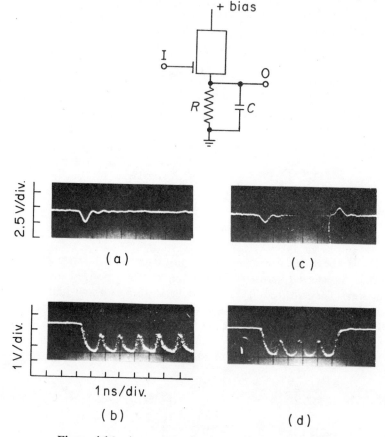

**Figure 4.16** A memory circuit, and its corresponding signals for $R = 50\,\Omega$, $C = 4\,\text{pF}$. ((a) input signal, (b) output signal, (c) input signal and erase pulse, (d) output, terminated by erase signal.)

requires, firstly, careful study of the results by other workers, and secondly, practical experience particularly regarding the essential condition of highest cleanliness.

A few relevant points should, however, be outlined here. Most advanced work regarding pulse-signal applications has been performed on planar structures (Figure 4.17, page 88) where a semi-insulating substrate carries a semiconducting layer which has been deposited by liquid or gaseous epitaxy. Ohmic-contact electrodes are then applied to the top surface of the semiconducting layer, which has to have as high a mobility and as good a peak-to-valley current ratio as possible. The diode geometry is produced by etching away some parts of the *n*-layer on the substrate, a process called mesa-etching. This is done by photolithographic methods,

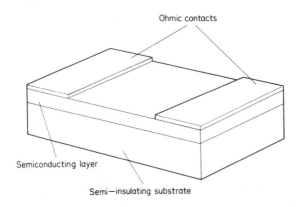

**Figure 4.17**   Planar structure.

which can be used for producing the most complicated device geometries. Heat is transferred across the substrate, which is bonded onto a metal heat sink. The substrate layer should be as thin as about half the inter-electrode distance along the *n*-layer. Much thinner substrate layers, although very attractive regarding improved heat sinking, are likely to

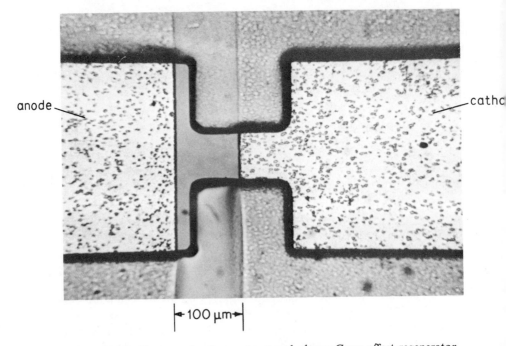

**Figure 4.18**   Photograph of mesastructured planar Gunn-effect regenerator.

cause electrical breakdown due to the fields between ohmic-contact electrode and the metal heat sink, which is usually earthed. There are numerous ways to achieve such a structure. One convenient and generally successful method is as follows. Anode contacts are produced by depositing $n^+$ liquid epitaxy layers. Anodes should preferably not be made of metallic materials, as metals migrate towards the cathode under the action of high electric fields and form a short-circuiting channel (see section 3.5). If a metal anode is chosen for convenience, the cross-sectional surface of

**Figure 4.19** Mesastructured diode with material channel shortening the active layer from anode to cathode, produced after a bias voltage of about three times the threshold was applied (the Au-tape used for bonding the electrodes into the strip line is clearly visible; the anode Au-tape seems to have contributed to the short-circuiting channel).

the active part should be increased at least two-fold in order to let the domains collapse before they reach the anode (see Figure 4.18). High-field anode edges should also be avoided on the semi-insulating surface which is, of course, revealed by the etch process for mesastructuring. Experimental results have indicated that metal can also migrate along the semi-insulating surface under the action of high fields.[10]

An example of anode material forming a channel along the semi-conducting surface is shown by Figure 4.19. It can be seen how the channel originated in the centre of the anode where the field is highest. Once this process of material migration has started, it seems to involve even the anode Au tape with which the electrodes are bonded into the strip line.

Ohmic cathode contacts are prepared by evaporating In–Ge–Ag and alloying at 600 °C for two minutes. Cathode edges should be very sharp in order to obtain good domain nucleating regions where the bias field is high. A similar metallization is used for preparing electrodes on the

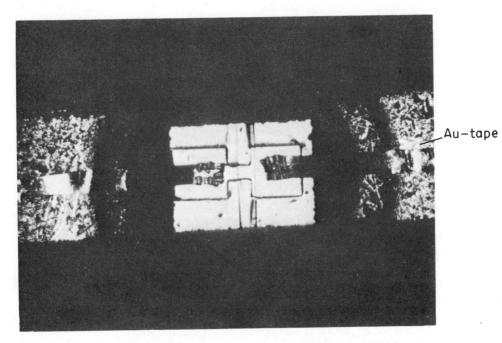

Au—tape

**Figure 4.20**  Mesastructure, deposited on a heat sink, is bonded via Au-tapes into the strip line or some other suitable circuit structure.

back part of the $n^+$ anode layers. The back face of the substrate is then polished and covered with Au–Ge. This is soldered at a temperature of about 330 °C with a Au–Ge preform onto a Au-plated header, acting as heat sink. The total assembly is bonded by using Au-tapes into a strip line of suitable characteristic impedance (see Figure 4.20). The thermal resistance of the resulting devices describes the ratio of temperature difference between the hottest part of the $n$-layer and the heat sink, and of the power dissipated. Under CW conditions, a 1 Watt bias power can easily generate about 500 mV pulse-signal amplitudes into a 50 Ω load, and the thermal resistance (see next section) would then be around 150 °C/W. This is, of course, no basic limit, and by suitable mesastructuring techniques one can etch out several 100 μm-wide *n*-layer strips in parallel on the same substrate for optimum heat-sinking.[58] In fact, it has been shown with pulsed bias operation that a total width of 2.5 mm for 10 μm-thick material can still produce good domain pulses and yet does not represent the ultimate limit.[13] In this case, the pulse-signal output power into 50 Ω loads was 4 W, whereas the CW output power could be expected to be around 1.5 W.

CW-operated pulse-signal devices have now been produced by several laboratories and have been operated over extended periods, wherever the fabrication parameters were chosen correctly, and good life-times can be expected.

For Schottky-gate devices a suitable Schottky-gate metal is deposited, again by evaporation or by sputtering. The techniques described here are, of course, very suitable for monolithic integration which should be the ultimate aim when larger commercial systems are developed.

## 4.5  THE THERMAL RESISTANCE

All CW-operated semiconductor devices require good thermal heat sinking for optimum operation. A convenient measure is the thermal resistance,[11] giving the ratio of maximum temperature increase and power dissipated by the device. This is primarily determined by the thermal conductivity $\kappa$ of the semiconductor and the thermal resistance of the semiconductor-metal interfaces. $\kappa$ is strongly temperature-dependent.[32]

The temperature increase for Gunn-effect pulse regenerators has to be kept as low as possible due to the decrease in peak-to-valley current ratio for increasing temperatures (see Figure 2.1, page 6). For such applications as pulse communication high signal powers are aimed at, and with the maximum efficiency being relatively low, the power dissipated in such devices is high. Therefore the thermal resistance has to be kept as low as possible.

Unfortunately planar technology with semi-insulating GaAs substrates introduces a basic limitation of thermal resistances, because the heat has to be dissipated across the low conductivity substrate. One could argue that a sandwich-type construction with the active layer between two parallel ohmic-contact surfaces would be of advantage because heat could be drained out via the metallized surfaces directly. However it is difficult and ultimately impossible to fabricate I.C.-structures if the inter-electrode distance has to be much more than 10 μm. Therefore practically all important results on pulse-regenerating d.c.-operating Gunn-effect devices have been obtained with planar structures.

Thermal resistances of active devices can be determined by measuring the change of some material parameter with temperature and thus deriving the temperature increase for a given power applied. Often thermal resistances are determined by measuring the increase of electrical resistance with temperature. Firstly the device is heated externally and the resistance changes are recorded. Subsequently the d.c.-operating voltage is applied and causes a resistance change due to heating. Usually the two measurements do not correspond to each other exactly as the temperature distribution due to the d.c. bias is not as uniform as that due to external heating. However, in planar structures in contrast to, say, sandwich devices, the heat distribution is more uniform.

The *I–V* curves at various ambient temperatures, as shown by Figure 4.21, are obtained with a pulsed monitoring voltage of very short duration and low repetition frequency. The d.c. *I–V* curve with the heat sink at room temperature is then measured. The cross-points give the device temperatures for the d.c. voltages applied, and the resulting thermal resistance is shown by Figure 4.22 for voltages up to the threshold for the transferred-electron effect.

The thermal conductivity for semi-insulating GaAs can be expressed approximately by

$$\kappa = \kappa_0 - \kappa_1 T \qquad (4.3)$$

where $\kappa_0 = 0.76$ W/K cm,
$\quad \kappa_1 = 10^{-3}$ W/K$^2$ cm,
$\quad T$ is the temperature in K.

After solving the differential equation for heat flow, the temperature distribution across the substrate is obtained by the following expression, where the distance variable $x$ starts at the metal-substrate interface:

$$x = \frac{A_T \theta_T}{(T_1 - T_s)}\left[\kappa_0(T - T_s) - \frac{\kappa_1}{2}(T^2 - T_s^2)\right] \qquad (4.4)$$

where $T_s$ and $T_1$ are the temperatures at $x = 0$ and at the interface $n$-layer to the substrate, which is at $x = d_s$, and $A_T$ is the cross-sectional surface of heat flow.

The thermal resistance is correspondingly

$$\theta_T = \frac{d_s}{A_T[\kappa_0 - \frac{1}{2}\kappa_1(T_1 + T_s)]} \qquad (4.5)$$

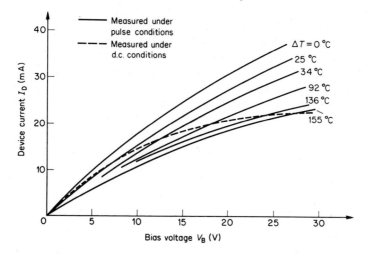

**Figure 4.21** Device current $I_D$ *vs.* bias voltage $V_B$.

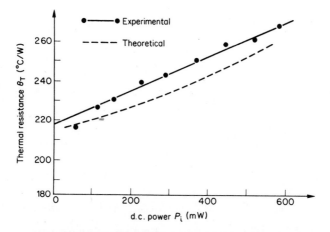

**Figure 4.22** Experimental and theoretical thermal resistance $\theta_T$ *vs.* d.c. power $P_i$ supplied to the n-layer (for $P_i = 600$ mW the n-layer temperature is 240 °C).

which agrees well with experimental results, as shown for the case of Figure 4.22. A difference between experimental and theoretical $\theta_T$ can be attributed to thermal resistance at the metal-substrate interface. In fact, it can be seen that the heat-sinking technique employed produces a satisfactory heat flow across the metal-substrate interface.

A more exact treatment of thermal resistances which also includes those of mesastructuring where lateral spreading of the heat-flow lines in the substrate is possible, can be found in reference 27. It is, of course, of great advantage to have thermal spreading in a large substrate, as $\theta_T$ can be considerably reduced. In fact by careful diode structuring, several relatively narrow active layers can be situated on one substrate in parallel, thus producing a minimum value of $\theta_T$.

Another possibility of reducing $\theta_T$ is reducing the substrate thickness $d_s$. Unfortunately there is a limit imposed by the electrical breakdown across $d_s$ due to the application of a high field. Such breakdown causes permanent damage as it usually short-circuits the ohmic electrode, carrying a high d.c. voltage, with the heat-sinking metal base which is generally earthed. Practical breakdown fields are around $10^4$ V/cm for good substrate materials. This means a 1 gigapulse device with $l = 100$ μm has to have $d_s$ at least 50 μm thick, or, $d_s > \frac{1}{2}l$.

If a constant output-signal power for all values of $l$ is considered, then the device current near the threshold field has to be taken, namely

$$I = \frac{E_t A}{\rho} \tag{4.6}$$

This is indeed independent of $l$. The power dissipated in the Gunn-effect layer for this current is

$$P = ne\mu l A E_t^2 \tag{4.7}$$

Inserting this expression for P into equation (4.5) by replacing $\theta_T$, one finds

$$\frac{T_1 - T_s}{ne\mu l A E_t^2} = \frac{d_s}{A_T[\kappa_0 - \frac{1}{2}\kappa_1(T_1 - T_s)]}$$

With $d_s \approx \frac{1}{2}l$, $A = wd$ and $A_T = lw$, one obtains

$$(nl)e\mu E_t^2 d = 2(T_1 - T_s)[\kappa_0 - \frac{1}{2}\kappa_1(T_1 - T_s)] \tag{4.8}$$

As the terms on the left-hand side of the equation are constant, it can be concluded that $T_1$ is independent of $l$ and thus of pulse rates. The values of $T_1$ will therefore always be relatively high (see Figure 4.22) unless thermal spreading is widely employed. In addition, some heat sinking occurs across the contacts and causes a decrease in $T_1$. The practical values of $T_1$ obtained by various workers so far for planar Gunn-effect regenerators are therefore around 150 °C,[11,58] and can possibly be reduced by decreasing $w$ and paralleling several narrow layers on one substrate for good heat-flow spreading.

The exact details of the mechanism of breakdown are not yet entirely clear. It has been found that semi-insulating GaAs shows an increasing current with rising applied voltage. This is caused by traps being filled from a certain level of carrier injection. Before these traps are filled, the current remains well below the value corresponding to space-charge limitation. At the point of trap filling there is a rapid increase in current with applied voltage until the current reaches the limit due to space charge limitation. For many types of dopant (e.g. oxygen) the level of trap-filled voltages are relatively low, so that substrates made with such dopants are not suitable. Fortunately Cr-doped semi-insulating GaAs shows a behaviour[18] which indicates that the trap-filled voltage level occurs at very increased values. The Fermi level then seems to be pinned closely to the deep Cr-level, resulting in a high concentration of traps, and a very large number of charge carriers would have to be injected to fill the traps. In practice, a different kind of breakdown occurs first which is destructive and might have some avalanching effect.

### 4.6  APPENDIX

The figure of trigger capability is defined for two-terminal devices as

$$\gamma_A = \frac{\Delta I_D}{\Delta V_{Amin}} = \frac{\Delta I_D}{\Delta I_{min} R_0} \cong \frac{k' E_t}{\Delta E_{min} R_0}$$

where $\Delta I_D$ is the current drop due to domain nucleation (proportional
to $E_t - E_R$),
$\Delta I_{min}$ and $\Delta E_{min}$ are the minimum current and field increase
respectively required for domain nucleation,
$\Delta V_{Amin} = \Delta I_{min} R_0$.

Correspondingly the figure of trigger capability for a Schottky-gate
device is

$$\gamma_G = \frac{\Delta I_D}{\Delta V_{Gmin}} = \frac{\Delta I_D}{\Delta E_G} \cdot \frac{\Delta E_G}{\Delta V_{Gmin}} = \frac{\Delta I_D}{\Delta I_{min}} \cdot \frac{A}{\rho} \cdot h_g$$

where $\Delta V_{Gmin}$ is the minimum voltage signal to be applied to the gate
for successful domain nucleation,
$A$ is the cross-sectional surface of the semiconductor layer,
$\rho$ is the low-field resistivity,
$h_g$ is either determined computationally or by approximation, and
gives the ratio of field increase underneath the gate and the voltage
difference applied to it.
Using the above expressions one finds

$$\gamma = \frac{\gamma_G}{\gamma_A} = l_t h_g$$

which is equation (4.2) for a simple approximation of $h_g$.

# 5:  Logic Systems

Once logic elements are available it is necessary to show that they can be usefully combined to form a total system. In particular, as Gunn-effect domain devices are envisaged for ultrafast pulse applications, numerous important aspects have to be clarified, such as stability, speed of the whole system, and therefore also the packaging densities possible. Of course, there will be applications where individual devices are employed with long distances between them, such as the reshaping of pulses in pulse communication systems (see Chapter 6), and the usefulness of the new devices does not therefore exclusively depend on packaging.

So far, no logic system of any size has yet been developed. However, several smaller units such as a shift register and an A/D (analogue-to-digital) converter have been operated in laboratories. This demonstrates that systems can be made to function correctly and applications can now be envisaged.

A limited amount of success has been obtained when employing two-terminal devices (see section 4.2). The difficulty here is that domains generate signals which travel in both the directions of input and output line, and therefore special arrangements have to be made to ensure stable operation. Three-terminal devices (section 4.3) are better here and impressive results have been achieved with Schottky-gate electrodes.

## 5.1  SHIFT REGISTER WITH TWO-TERMINAL DEVICES

A shift register consists of a chain of memory units together with some associated logic circuitry, and binary digits can be added and stored. In this section we consider the following circuit possibility for which we describe also some first experimental evidence that has been obtained so far.

Pulses are applied to an input terminal of the register chain consisting of logic units which operate as follows. When a first input pulse enters such a unit, it is stored and no output pulse occurs. A subsequent input pulse erases the stored signal and produces an output pulse which is then applied to the next unit. The pulses stored in the units of the chain therefore represent the added sum of all input pulses applied to the shift register. That means the last digit of the binary sum is stored in the first unit, the second to last digit in the second unit, and so on. Such a chain is very useful as a counting circuit for binary digits, and its high operational speed is attractive because very fast pulse signals occurring at gigapulse rates can be added up for evaluation purposes. Examples are nuclear or photon counters. As the delay times between the units is of no importance for the operation of such a shift register, this particular application does not have to rely on high packaging densities. Such a circuit could therefore possibly represent the first commercial application of a Gunn-effect logic system for signal processing, other than pulse communication (see Chapter 7).

A unit can be set up (Figure 5.1) by using a Gunn-effect AND1, whose ouput is connected to a short delay line. One of the input terminals of this gate is connected to another Gunn-effect diode $D_G$ which is biased just above threshold and therefore generates domain pulses continuously.

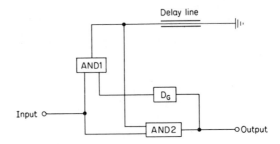

**Figure 5.1**   Basic unit of Gunn-effect shift register.

These pulses raise the field in the diode of AND1, but domains are only nucleated by the simultaneous application of a signal to the other input terminal of AND1. The resulting domain pulse then travels along the delay line, is reflected at the end, and is applied to the diode of AND1 again, resulting in another domain nucleation if pulses arrive simultaneously from $D_G$. A second input signal succeeds in nucleating a domain in AND2 because signals from AND1 are applied simultaneously to its second input terminal. The output signal from AND2 interrupts domain nucleation in $D_G$ by lowering the potential below threshold, and the stored signal in the delay line does not renucleate a domain in AND1 and is therefore erased. The output from AND2 is directly connected to the input of the subsequent unit of the shift register.

The operation of such a unit was obtained as follows. A bias pulse was applied to diode $D_G$ (see Figure 5.2), where continuous domain nucleation was produced. The resulting domain signals were applied to another Gunn-effect device acting as AND1. Using attenuators, these signals were just prevented from nucleating domains in AND1. A differentiator was used to produce a trigger pulse from the starting slope of the bias pulse in order to simulate a positive input signal. This trigger pulse was applied to the other terminal of AND1 and succeeded, together with the domain signals from $D_G$, in nucleating one single domain there. The resulting domain pulse was applied to a delay line. When this pulse returned to the diode of AND1, a new domain was nucleated as long as signals arrived

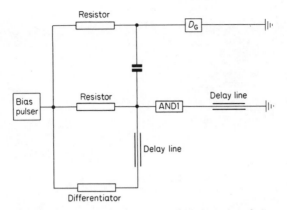

**Figure 5.2** Circuit of first feasibility study of Gunn-effect shift register.

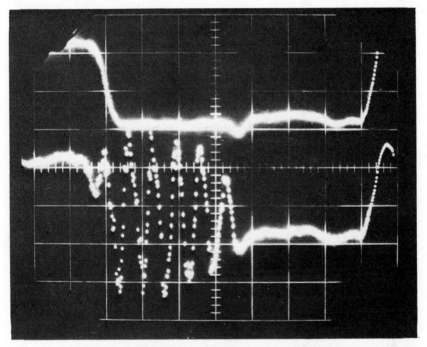

**Figure 5.3** Output from AND1 (top) and differentiator (bottom) (vertical scale, 50 mV/div.; horizontal scale, 5 ns/div.).

simultaneously from $D_G$. The frequencies of the signals produced by AND1 are therefore not given by the domain transit time but by twice the signal transit along the delay line. This operation is shown in Figure 5.3, where the lower curve shows the pulse from the differentiator,

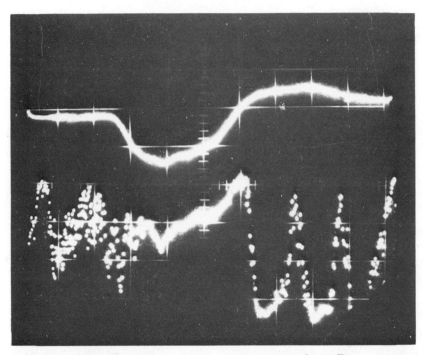

**Figure 5.4** Input to $D_G$ (top), output from $D_G$ (bottom) (vertical scale, 200 mV/div.; horizontal scale, 1 ns/div.).

whereas the upper curve shows the resulting delay-line transit wave of AND1. By lowering the bias-pulse amplitude after the application of the signal from the differentiator, thus simulating the effect of a second input signal being applied to a gate AND2, the signals from $D_G$ were interrupted (Figure 5.4), and the signal stored in the delay line was erased. This performance is shown in Figure 5.5, where the upper trace (i.e. output from AND1) shows delay-line transit oscillations starting when the signal arrives from the differentiator (middle trace), and terminating when the signals from $D_G$ are interrupted (bottom trace).

Future development work has to produce such a unit in integrated circuit form in order to make the dimensions smaller than a wavelength of the basic pulse frequency. This will then avoid difficulties due to internal reflections causing unwanted resonances. In order to avoid signals travelling from one unit to another in the wrong direction, or when no information transfer is desired, the following arrangement of fast two-electrode directional gates has to be set up.

Three regenerator diodes are set up in series. Each of them is supplied with a d.c. bias voltage $V_B$ well below threshold, and an individual synchronization pulse $S_p$ which raises the diode field to a value just below

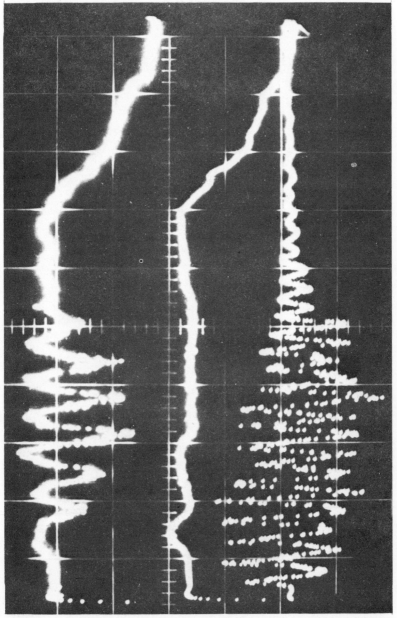

**Figure 5.5** Output from AND1 (top), output from differentiator (middle) and output from $D_G$ (bottom) (vertical scale for 2 upper traces, 50 mV/div., for lower trace, 200 mV/div.; horizontal scale, 5 ns/div.).

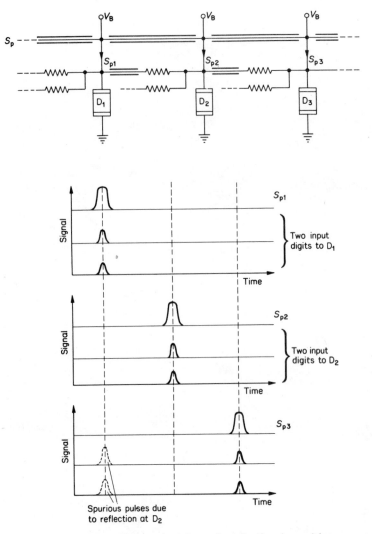

**Figure 5.6** The principles of reflection-insensitive three-element regenerator. Spurious pulse signals (see bottom) do not succeed in nucleating a domain in $D_1$ because there is no $S_{P1}$ available.

threshold (see Figure 5.6). $S_p$ is applied to the first regenerator only when a new digit is put into the shift register. If a pulse then has to be transferred, it can be employed to nucleate a domain in this first regenerator, whereas the subsequent two diodes are still unable to support a domain. When this domain has grown sufficiently, a pulse $S_p$ is also applied to the second regenerator so that the signal can now proceed one step further. When the domain and $S_p$ of the first regenerator have disappeared, regenerator 3 is energized by an $S_p$ pulse and a domain can now be nucleated there too. A domain signal can thus be ferried through a system which acts like a lock for lifting ships, and if the sequence of $S_p$ pulses applied to the three regenerators is correctly chosen, no pulse signal can travel in the reverse direction. The time delay required is at least slightly more than a domain transit. The shift register can be envisaged to operate with a digital repetition frequency of a little more than a domain transit time. This locking system therefore only reduces the read-in speed of the regenerators by a small amount, but the disadvantage is additional sophistication and complexity of the system.

## 5.2  SHIFT  REGISTER  BASED  ON  THREE-TERMINAL DEVICES

The difficulties arising in connection with instabilities are considerably reduced if three-terminal devices are employed, as described in section 4.3. Indeed, some impressive results have been achieved by using Schottky-gate trigger devices[57] (see page 82).

A successful development is represented by a feasibility study of a shift register based on the memory element of Figure 4.16, page 87. The basic unit of such a shift register is explained by Figure 5.7. A negative input signal applied together with a shift pulse to the gate electrodes starts continuous domain nucleation. When a subsequent shift signal appears, domain nucleation is obtained in the second Gunn-effect device.

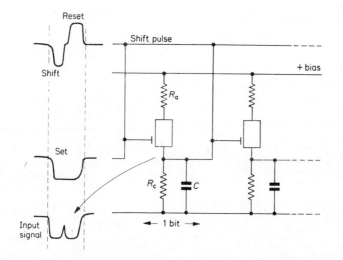

**Figure 5.7**   Basic unit of shift register.

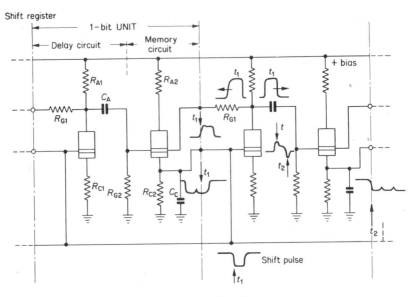

**Figure 5.8** Shift register.

A positive pulse, acting as reset signal, can be applied to the line connected to the anode electrode and prevents further domain nucleation. Of course, this circuit does not yet function as a proper shift register and a further delay element has to be introduced so that the final system is like that of Figure 5.8. When the first stage stores a signal in its memory unit, a negative shift pulse applied to the lower line of Figure 5.8, and the negative pulse from the anode of the first memory unit, together succeed in nucleating a domain in the diode of the subsequent delay unit. A positive pulse is obtained at the anode of this delay unit and is applied both to the gate of the preceding memory circuit to extinguish the memorized state of the first stage, and to a differentiating capacitance circuit. The differentiated signal contains a positive pulse followed by a negative one, and the latter pulse nucleates a domain in the second memory unit when applied to the gate electrode. In this way, the pulse signal is shifted from the first unit to the subsequent one, and shifting occurs by one step each time a shift signal is obtained. Between each digital signal applied a shift pulse is therefore required, and the stored pulses are exactly as they arrived, without being added up, as was the case with the shift register described in section 5.1. It is, of course, also possible to develop a Schottky-gate Gunn-effect shift register which adds up the binary signals applied. However, the present circuit represents a first proposal. The results obtained so far demonstrate that the unidirective property of Schottky-gate devices avoids the difficulties of reflections and resonances that are encountered with two-terminal device systems.

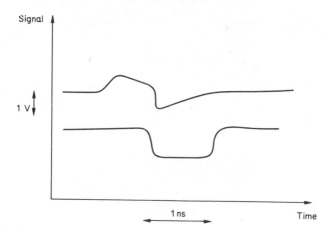

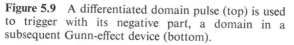

**Figure 5.9** A differentiated domain pulse (top) is used to trigger with its negative part, a domain in a subsequent Gunn-effect device (bottom).

So far the nucleation of a domain in one Gunn-effect device (Figure 5.9, bottom) by the negative spike of a differentiated domain pulse from the anode of another Gunn diode (Figure 5.9, top) has been verified, together with the correct functioning of the circuit of Figure 5.7, as shown by the waveforms of Figure 5.10. Here (a) shows the set and reset symbols, (b) shows the shift pulse, (c) the signals from the first memory unit, and (d) the waveform of the second memory. After this first feasibility study it should now be possible to set up a total system as proposed by Figure 5.8. Again, a realization in monolithic integration would be advisable in order to exploit the speed advantage.

## 5.3 ANALOGUE-TO-DIGITAL CONVERTER

An analogue-to-digital conversion can be achieved by either a travelling-signal detection or a triangular-bias pulse technique. The first method arranges for the analogue signal to travel along a chain of intermediate attenuators and delay sections. After each attenuator, the analogue voltage is applied to a fast Gunn-effect diode biased just below threshold. If the attenuated signal is large enough, a domain is nucleated. The first $n$ diodes will therefore produce a domain pulse representing $n$ quanta of the quantized analogue voltage. These pulses have to be added up by a suitable adder of the type described in section 5.4.

The other technique employs an ordinary Gunn-effect diode, which is supplied with a triangular bias pulse together with the analogue voltage. The analogue voltage raises the tip of the triangular pulse above the threshold value. Depending on the amplitude of the analogue voltage, the diode will therefore be biased above threshold for the time of a given number $k$

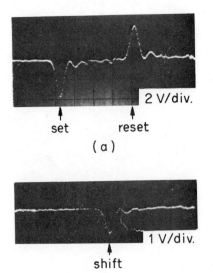

2 V/div.

set      reset

(a)

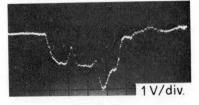

1 V/div.

shift

(b)

1 V/div.

(c)

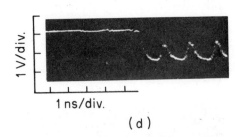

1 V/div.

1 ns/div.

(d)

**Figure 5.10** Waveforms of shift-register unit.

of the domain transits. The diode will then show $k$ output pulses, which are the quanta of the quantized analogue voltage. These pulses have to be added up again in the way outlined later, in order to produce a suitable coding. The required triangular pulse can be produced by a suitably-shaped Gunn-effect diode, as the output-pulse shape of Gunn-effect

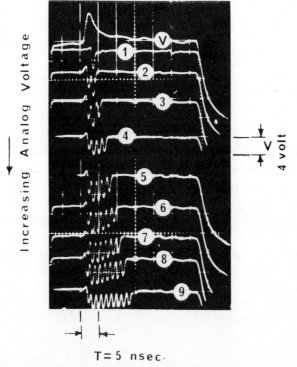

**Figure 5.11** Triangular pulses produced by device shaping.

diodes is determined by a variation of its cross-sectional surface or its conductivity profile between the electrodes. Figure 5.11 shows the production of triangular pulses (resistivity $1.95\ \Omega$ cm, $n = 3.7 \times 10^{14}/cm^3$, mobility $5300\ cm^2/Vs$, $l = 0.5$ mm, low-field resistance $350\ \Omega$). The output from an A/D converter of the latter type is shown in Figure 5.12, where a quantization of 9 has been obtained. Further work has in fact produced quantization of up to 17 without employing sophisticated stabilizing equipment. In order to obtain pulse coding, some suitable logic circuit has to be employed.

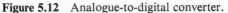

**Figure 5.12** Analogue-to-digital converter.

## 5.4. GUNN-EFFECT ADDER AND OTHER CIRCUITS

Several proposals of other circuits have been advanced. However no feasibility study has yet been performed, but as soon as three-terminal regenerators have become more generally available for d.c. biasing, a sequential adder will probably be one of the first to be set up because it will be of use for numerous applications.

A sequential Gunn-effect adder was first discussed in 1968[25] when, as a first feasibility approach, blocking capacitors were employed in order to avoid difficulties with the required bias polarities at various points in the circuit. Ultimately, of course, the system has to be designed without these capacitors, because they limit the frequency of operation. The circuit was proposed along classical lines, using the logic elements described in the last chapter. The two input binary digits are simultaneously applied to an 'exclusive OR' (see Figure 4.12) and to an AND. The former produces the 'SUM' whereas the latter gives the 'CARRY'. Both have to be added after the carry has been delayed by one digital time-spacing. This is again performed by a second 'exclusive OR'. If an output occurs from the first 'exclusive OR' at the same time as a signal emerges from the delay element, this second summation produces a carry which has to be added, after a further delay, to any subsequent signal from the first exclusive OR. This second carry is generated by a second AND. As the emergence of a signal from the first exclusive OR means that one time-spacing later no signal emerges from the first delay element, the output from the second AND can be added again to the input of the second exclusive OR, in parallel with the first AND. The total circuit consists, therefore, of only two exclusive ORs, two ANDs, two delay elements and, if single-chip two-electrode exclusive ORs (see section 4.2 and Figure 4.3, page 73) are employed, several pulse-signal-polarity reversing series regenerators. One can expect here that instabilities are very likely, unless some directionality is introduced for the signal flow.

## 5.5 PACKAGING DENSITIES AND ULTIMATE SPEED LIMITATIONS

In connection with logic systems, the high speed of operation possible with Gunn-effect devices can only be fully exploited if high packaging densities can be achieved. This is required in order that signals can travel from one logic gate to another within times which are shorter than the pulse-processing delays. For increased pulse rates the distances between Gunn-effect devices need to be extremely short so that the power densities produced by a total logic system become too high for a satisfactory removal of the waste heat. Operation will then become impossible as the system will soon become overheated. For very fast systems one might have to consider new methods of cooling, such as forced air or liquid cooling or evaporation cooling along similar lines to a heat pipe, or the logic system might ultimately have to be immersed in liquid nitrogen so that the device temperatures are then at reasonable values somewhere around 100 °C.

The power dissipation of a Gunn-effect diode for logic applications, that is biased near threshold, is given by

$$P \simeq E_t^2 A e \mu n l$$

With $nl = 10^{13}/\text{cm}^2$ for the formation of mature domains, and with $nd = 2.10^{11}/\text{cm}^2$ as determined in Chapter 3, one finds

$$P = 40 \, \frac{l^2}{\text{cm}^2} \, \text{Watts}$$

or, for 1 gigapulse/second, $P = 4 \, \text{mW}$ and for 40 gigapulse/second, $P = 2.5 \, \mu\text{W}$. Assuming that interconnections are one-fifth of the wavelength $\lambda$ of the repetition frequency at the pulse rate of operation, the volume associated with each active device would be about $\frac{4}{3}\pi(\lambda/5)^3$. Here the wavelength would have to be that of the transmission line, which could well be a highly conducting strip on a semi-insulating GaAs dielectric the relative permittivity of which is 12. The waste-power density of a monolithic GaAs system (including GaAs strip lines) would then be

$$\frac{1}{8} \, \text{mW/cm}^3 \text{ for 1 gigapulse/second}$$

and $\qquad$ $5 \, \text{mW/cm}^3 \text{ for 40 gigapulse/second}$

Another attempt at assessing the speed limitations of pulse devices is to find the power-speed product. This gives the energy lost during the switching time $T_D$. The best Si-junction logic devices at present give 20 pJ and the technological limit has been estimated to be around 1 pJ. For Gunn-effect devices this figure can be estimated as follows. The waste power is again $P = V_t^2/R_0$ and the switching time for optimum domain devices is $T_a \simeq \frac{1}{9}\tau$ (see equation 2.66). The product of both terms is then

$$PT_a = \tfrac{1}{9}\varepsilon E_t^2 \times \text{Vol.}$$

(Vol. = the volume of the active device). Taking the smallest value of Vol. possible for 40 gigapulse/second as given by the limiting $nd$ and $nl$ products, one would find

$$PT_a \simeq 10^{-16} \, \text{J}$$

This value is, of course, only acceptable if reasonable device impedances $R_0$ and carrier densities are still possible. In fact, by keeping $l = 2 \, \mu\text{m}$, $d = 3l$, $n = 10^{16}/\text{cm}^3$ (i.e. $\rho \simeq 0.1 \, \Omega \, \text{cm}$) and $R_0 = 100 \, \Omega$, the same value is obtained. Even if one feels happier in not taking the very limiting values of *l* and *d*, one still obtains a power-speed product which is impressively shorter than that possible with junction devices.

Of course, an important speed limitation is given by the electron-transfer effect. Detailed computations have shown that the negative-differential mobility no longer exists at 100 GHz, due to scattering times limiting the speed with which electrons are transferred to the satellite valleys. Another limitation occurs due to the filling up of the central valley with electrons if the carrier density *n* is increased. For higher pulse rates, *l* has to be decreased. In order to maintain the critical *nl* product (i.e. it is advisable to have $nl \geqslant 2 \times 10^{12}/\text{cm}^2$ for pulse applications, where mature domains have to be obtained), *n* has to be increased correspondingly. For 40 gigapulse/second, *n* becomes about $10^{16}/\text{cm}^3$, which approaches the limit for the negative differential mobility still to be observed. It is therefore estimated that the highest pulse rates which can be envisaged are around 40 gigapulse/second. However, this speed limit is included in the $PT_a$ value derived above.

A clear disadvantage of Gunn-effect pulse devices with respect to field-effect transistors (e.g. MESFETs, for MEtal Semiconductor FETs) regarding high packaging densities, is the high current flowing when no signal exists.[36,37] In other words, it is not possible to have a similar set-up with Gunn-effect devices as with 'Normally off' FETs. It is here that some combination of both types of fast pulse devices, GaAs MESFETs and Gunn-effect elements, will be useful. Both components are certainly complementary to each other in many future applications.

MESFETs have a Schottky-junction gate, and the highest frequency of operation with any transistor has been achieved with GaAs MESFETs. The frequency and switching time is limited by the delay of removing charges away from the gate region. Therefore the gate width has been reduced considerably and values of around 1 μm have been obtained by photolithography. It has been shown by computation,[48] that the best conditions can be found with a width of 0.1 μm, although the gain will be reduced considerably already. Such widths can only be fabricated with electron-beam techniques for the exposure of a suitable resist, as optical wavelengths impose a limit of about 1 μm. The alignment problems for the two sets of masks required regarding the two different metallization steps of the Schottky-gate material and the ohmic contacts of source and drain, are severe, and were partly solved by first using Ni for both the Schottky electrode and the ohmic contact areas with a suitable first mask, which enabled the deposition of Ni on all contact areas. Au–Ge was subsequently deposited on the ohmic contact areas only by using a second mask where the edges of source and drain were kept a few micrometres back with respect to the Ni edges. In this way a small misalignment of this second mask did not matter. When the ohmic contact metals were then heated for alloying, Au–Ge ran to the source and drain edges automatically because Ni acts as a wetting material.

Unfortunately the removal of electrons under the gate cannot take place with saturation velocities, as the fringing fields of the gate electrode saturate the current first so that saturation velocities are never reached underneath the gate. Therefore the shortest pulse durations handled by MESFETs so far are around 500 ps, and it is possible that this value can only be decreased by a small amount. Schottky-gate Gunn-effect regenerators, on the other hand, have the advantage that the instability effect aids switching so that higher current levels can be switched faster. Additionally, the trigger gate does not have to extend over the whole width of the device because transverse growth of a Gunn-effect domain is extremely fast (approaching the velocity of light[40,62]). Therefore the gate electrode can be made to have exactly the length which gives the desired input impedance for matching. In fact the gate width does not have to be so excessively narrow as with the MESFET as the fringing field is employed for domain triggering. A final advantage is often given by the fact that Gunn-effect regenerators are usually monostable devices of simple construction, whereas transistor monostable multivibrators require additional sophisticated circuitry. On the other hand, a disadvantage of Gunn-effect devices is the requirement for the bias voltages to be highly stabilized in order to achieve a useful value of pulse-regeneration gain. However, as both types of elements are produced with the same materials, integration is easily possible by incorporating Gunn-effect elements and MESFETs. Here the new developments with liquid epitaxy are valuable, where also thin, high-conductivity layers for FETs have been grown by dipping a cold substrate into the heated GaAs-saturated melt of Ga or Sn.

## 5.6 APPLICATIONS

It is, of course, still early to estimate the applications which will ultimately become commercially attractive. However, it is useful to consider here the major areas of applications where Gunn-effect logic will represent definite advantages.

Firstly, base-band, gigapulse-rate communication systems can usefully employ Gunn-effect logic, and owing to their importance some further details are especially discussed in the next chapter. Then those aspects of digital instrumentation might benefit greatly where ultrafast speed is required. Examples are fast counters of individual pulse signals, sampling systems operating at high speed, and fast data processing. Counters would be especially useful for various scientific purposes where useful experiments cannot be performed, as no suitable counters are available yet. Here nuclear or phonon counters are possibilities. The circuit required would have to be similar to the shift register described in this chapter.

Finally, it might be advantageous to build a central processor of a large digital system in ultrafast logic. Although good improvements have been achieved here recently by paralleling techniques, further advantages can be gained regarding the computational speed of such systems by increasing the speed of their central processors.[38] However, this last application still requires considerable attention, particularly regarding GaAs-materials quality and relevant I.C. technology, as a commercial processor would have to operate very reliably before a large, expensive computer could be made to rely on it.

# 6: Pulse Communication with Gunn-effect Devices

For numerous communication applications there exists a requirement for higher pulse rates, as this permits for further increased bandwidths the use, for example, of P.C.M. with its well-known advantages regarding noise immunity.[4] Unfortunately, it does not seem possible to use bipolar junction devices above 500 megapulse/second, as the junction capacitance has to be charged via the characteristic impedance of the line, and as the minority-carrier removal is a long process when switching into the reverse direction is performed. These effects involve a time constant which is longer than the periods of the desired gigapulse rates. Therefore one has to look for other effects which enable the development of devices suitable for very fast pulse-processing. Several phenomena have been considered, such as the tunnel effect, the switching of amorphous materials, and the Gunn-effect domain. Only the last possibility is considered to be suitable, because the tunnel effect does not give large enough output powers, whilst switching of amorphous materials does not operate reliably at high repetition rates over long periods. It must, however, be stated that Gunn-effect pulse devices require further work before they can be incorporated in new microwave-bit-rate communication systems.

This chapter outlines the present state of development of Gunn-effect pulse communication and discusses the possibilities of using the resulting base-band pulse signals to modulate modern microwave semiconductor devices in order to obtain frequency, phase and amplitude shift-keyed systems (F.S.K., P.S.K., A.S.K.).

## 6.1 BASE-BAND BINARY COMMUNICATION

Pulse Code Modulation can be of advantage with respect to Analogue Modulation Techniques because of its noise immunity. This means also that fewer repeater stations are required, and this is valid for bandwidths of more than about 10 MHz in connection with power levels common with CW Gunn-effect diodes. Using the very high pulse rates available with transferred electron devices, it is certainly attractive to consider its use for binary communication.

The microwave bit-rate signals can be obtained either by using the analogue to digital converters described in the last chapter, or by time-division multiplexing lower pulse-rate channels. The latter can easily be achieved by using several Gunn-effect diodes together with timing devices such as step-recovery diodes.

Step-recovery diodes can be used for the production of sharp triggering spikes. When a sinusoidal voltage is applied together with a bias voltage, the junction capacitance is charged up during the period of forward current. When the reverse voltage is applied, the diode will only conduct until the whole junction has been depleted of charge. The reverse current then stops very suddenly. The resulting current step can be differentiated to give a sharp spike. The level of the bias voltage determines the exact position of the spike during the negative period of the sinusoidal voltage.

The output spikes from a step-recovery diode are then applied together with the pulse signals from one of the lower pulse-rate channels to a Gunn-effect diode pulse reshaper, as described in section 4.1. Only the simultaneous occurrence of the triggering spike and an input signal pulse nucleates a domain in the Gunn-effect diode. The output will, therefore, be shortened and precisely-timed signals. Step-recovery diodes have been used to generate output pulses of about 70 ps duration and 5 V amplitude. This would therefore be suitable for up to about 5 gigabit/second information flow-rates.

Each low bit-rate channel is made to drive a unit consisting of a Gunn-effect diode reshaper and a step-recovery diode, the bias voltage of which is set so that one triggering spike is produced at an exactly defined position during the period of the low-frequency pulse signals. The step-recovery diode could, of course, be fed with the fundamental repetition frequency $f_1$ extracted from the low-frequency pulses by using a filter and a resonator at $f_1$. The Gunn-effect diodes are biased in such a way that only the simultaneous occurrence of the pulsed signal of $f_1$ and the step-recovery spike produces a domain, whose output pulse is at the multiplexing frequency $f_h$. A module consisting of one step-recovery spike generator and a Gunn-effect diode is employed for each low bit-rate channel. The resulting signals contain the same pulse information but the pulse duration is greatly reduced. By suitable arrangment of the bias voltage for each step-recovery diode, the shortened pulses can be positioned in phase so that the pulse signals from each channel are allocated their well-defined individual phase space. By using isolators the output from each module is then added together, thus giving the multiplexed signal at $f_h$.

Demultiplexing can be achieved in the reverse manner. The signal at $f_h$ is applied to a number of the above modules in parallel. The step–recovery diodes are fed with a sinusoidal signal at $f_1$. The bias voltages are adjusted for each step-recovery diode so that spikes will be produced at phase positions which are representative of the particular low bit-rate channel. These spikes are fed to Gunn-effect diodes with suitable bias supplied so that only the spike and a pulse signal at $f_1$ together succeed in nucleating domains. The long pulse durations required for the demultiplexing signals can be obtained in two ways. The Gunn-effect diodes can either have very long interelectrode distances, or they can be employed to trigger fast transistor multivibrators. Monostable transistor multivibrators have been successfully triggered with input pulses as short as 100 ps, provided the total charge injected was sufficiently large. The multivibrators can have output-pulse times of any desired value. This result demonstrates also that individual short Gunn-effect diode pulses can be studied with low-frequency oscilloscopes by using such a multivibrator.

Instead of using a step-recovery module, the relaxation oscillator by Fisher[15] can be employed. It is basically a Gunn-effect oscillator with an inductance $L$ in series. The load $R_{LP}$ is parallel to $L$ and larger than $R_0$. The bias voltage $V_B$ is such that on a d.c. basis the GaAs is biased above threshold. As soon as a domain is nucleated, less current flows through the diode. The growth time of the domain is short enough for $L$ to be open circuited, and a negative spike occurs across $R_{LP}$ (assuming $V_B$ is positive). However, soon afterwards (after a time $T_1 = L/R_0$) $L$ represents a short gain, and the voltage across $R_{LP}$ reverts to zero. Then the domain reaches the anode and collapses. A positive voltage now occurs across $R_{LP}$. The amplitude is smaller, as the diode-terminal voltage is reduced, giving a smaller domain voltage. $L$ now has to be discharged, before a new domain is formed, via the parallel combination of $R_{LP}$ and $R_0$, giving a time constant

$$T_2 = L(R_0 + R_{LP})/R_0 R_{LP}$$

which is slower than $T_1$.

The circuit constants can be altered to give short spikes of amplitude and repetition frequency $f_r$ suitable for the pulse rates of the regenerator. As $f_r$ is also dependent on $V_B$, one can arrange for a phase-locking loop by using a pass-band filter for the fundamental frequency of the pulse signal, and a phase detector for comparing the relaxation oscillator output with the filter output, and controlling $V_B$ via a d.c. amplifier. In this way the retiming spikes are also phase-controlled with reference to the average phase of many signal pulses. As $f_r$ is about one-tenth of the domain transit frequency, this spike generator can be expected to operate at marginally higher information flow-rates than the step-recovery module.

Whereas the method of time-division multiplexing is relatively straightforward, when gigapulse rates are required the direct analogue-to-digital conversion is still not yet sufficiently clarified because no reliable gigapulse-rate adder has been developed yet, although there are proposals available (see section 5.4).

Decoding of a gigapulse-rate signal stream, obtained directly by A/D conversion, can be achieved by semiconventional techniques. The $p$ pulses representing one analogue value have to be differently amplified so that the pulse amplitudes are made proportional to $q + 1$ for the pulses representing the digits $2^q$. This means that the pulse representing $2^1$ is double the size of that of $2^0$ and the pulse of $2^2$ is three times as large.

The weighting is performed with Gunn-effect diodes of different cross-sectional diameters, as this is proportional to the pulse amplitude produced. The individual Gunn-effect diodes can be triggered with retiming spikes. The weighted pulses are applied to an integrating circuit the output of which is the reproduced analogue signal.

The gigapulse signals are transmitted along a suitable line. At certain intervals they have to be regenerated by Gunn-effect diode regenerators. Such a process involves both shaping and retiming of the pulses. Reshaping can be achieved by a Gunn-effect diode of suitable $n$ and $l$ as described in Chapter 4. When the diode is connected to a load resistor in series, the applied bias has to be a voltage source; when in parallel, the bias has to be basically a current source. In the first case a signal reversal occurs, in the second case the signal polarity remains the same. The input-signal polarity has to be such that it produces an increase in crystal field above the threshold value.

The retiming circuit can be set up with the help of a step-recovery diode module, as outlined above.

## 6.2   F.S.K., A.S.K., P.S.K.

It is often of advantage if the digital stream is used to modulate a carrier, and F.M. modulation is one possibility resulting in a technique termed frequency-shift-keying (F.S.K.). The bandwidth required is about twice the inverse of the base-band pulse width $\tau$ plus the frequency shift $\Delta f$, which is usually also equal to $1/\tau$. Another possibility is to produce an amplitude modulation (A.S.K.) of the carrier with the base-band pulses. Although this requires less bandwidth, the advantages of F.S.K., such as constant-amplitude properties, smaller sensitivity on fading, and reduced error probability for small signal-to-noise ratios, are lost. A further possibility is phase-shift-keying (P.S.K.) of the base-band signals. The bandwidth requirement is the same as with A.S.K.; if the phase reference has to be extracted from the signal, (i.e. we have a non-coherent P.S.K. system, such as differentially-coherent P.S.K. or phase comparison systems), the noise threshold characteristics are also equivalent to an A.S.K. system.[16] Better performance can, of course, be achieved by a coherent P.S.K. system, as the phase reference is supplied independently. The choice of any of these modulation systems has to be made by considering the effects of transmission media and other influences (e.g. effect of fading, aerial fluctuations, particularly with airborne antennas, interference). Each modulation system often presents certain advantages for given circumstances.

Microwave Gunn-effect diodes in low-$Q$ resonators or with resistive loading can be switched very quickly as the domain instability is so strong that it grows very fast, e.g. if the load is purely resistive, the growth occurs within a cycle. $X$-band Gunn-effect diodes can be expected to deliver a reasonable CW power level under low $Q$ conditions. Similarly, avalanche devices can be employed as microwave generators for advanced binary communication systems.

In the following, a Gunn-effect F.S.K. system is described. Base-band Gunn-effect diodes, as described in section 6.1, produce, with resistive loading, pulse signals which are applied to an $X$-band Gunn-effect diode in a low-$Q$ resonator whose output power is frequency-modulated by a change in bias voltage. As the base-band voltage pulses are applied together with a suitable d.c. bias voltage, the frequency shift of the oscillator output occurs at the same pulse rates as the base-band pulses. Instead of applying the signal voltage to the microwave diode directly, one can drive a microwave varactor diode with it, which is also inside the cavity and which frequency-modulates the oscillator by its voltage-dependent capacitance.

The resulting signal has the frequencies $f_1$ representing the 'mark' (or binary 1) and $f_2$ standing for the 'space' (i.e. the binary 0). This signal is then applied to a regenerator after it has been distorted by travelling along a transmission line. There are two ways. Either the F.S.K. signals are demodulated down to base band, and the regenerator of section 6.1 is employed with subsequent frequency-shift-keying; or the following regenerator can be used.[2]

The signal is first applied to a filter passing $f_1$ only. The output is applied to a Schottky barrier diode acting as an envelope detector and, using a frequency-locked Gunn-effect or Impatt oscillator at the base-band frequency $f_b$, the pulse repetition frequency $f_b$ is extracted and is then applied to a step-recovery diode module producing retiming signals. This module consists of two step-recovery diodes in opposition to each other. The resulting wave is shown as the lower trace of Figure 6.1. These retiming pulses consist of a positive one at some suitable point in time, and of a negative one which follows the positive pulse about half a cycle later. They are applied to a low-$Q$ Gunn-effect diode $X$-band oscillator of frequency $f_1$. The original F.S.K. input signal is also applied to this oscillator via a dispersive input loop, so that the voltage of the F.S.K. signal frequency corresponding to the base-band 'space' position does not trigger oscillations when both the d.c. bias voltage and the retiming signal are applied, whereas the voltage of the other signal frequency representing the 'mark' for the binary 1 succeeds in initiating oscillations.

If oscillations of the $X$-band diode are triggered, they will continue only until the negative pulse of the step-recovery module occurs, which acts as an oscillation-extinction pulse. One utilizes here a bias voltage hysteresis of Gunn-effect oscillators: once oscillations have been triggered the bias voltage can be reduced before oscillations are inhibited, because the r.f. voltage swing of the resonator brings the diode-field above the domain-nucleation threshold.

A first feasibility study was performed successfully so far. The output from a two-step-recovery diode module was applied to a Gunn-effect diode coaxial oscillator whose $Q$ was about 50. The bias voltages were such that each positive spike alone succeeded in nucleating oscillations. The resulting signal was applied to the second input of a very fast sampling oscilloscope, and the upper trace of Figure 6.1 shows the resulting envelope of the recorded microwaves. One can see how the retiming produced steep rise and fall times for the oscillations. Of course, the base-band frequency was still too low for utilization at microwave pulse rates, but the feasibility was demonstrated.

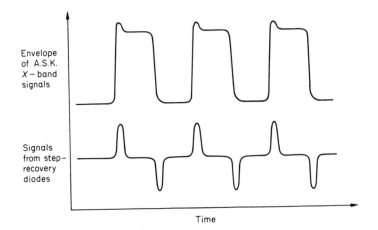

**Figure 6.1**  Wave functions of an A.S.K. modulator. (Top: envelope of A.S.K. signals at $X$-band. Bottom: trigger signals from a module with two step-recovery diodes.)

The output from this oscillator is applied to a second oscillator, operating at a free-running frequency $f_2$. When a burst of oscillations of frequency $f_1$ arrives from the first resonator, this second one is frequency-locked to $f_1$. The output from it is then again the frequency-modulated base-band signal, which has, however, now been retimed and reshaped.

In order to investigate the aspects of the triggering thresholds for the above regenerator oscillator, the following studies were performed.[33] A trigger signal of frequency $f_t$ and power $P_i$ as absorbed by the oscillator input (i.e. resonator + diode), was applied to a coaxial Gunn-effect oscillator, the natural free-running frequency of which was $f_0$. The input power was provided by a klystron, and was applied via a variable attenuator to the resonator using an inductive loop. A directional coupler permitted application of the signals originating from the oscillator (i.e. the output of the oscillator plus the reflected part of the trigger signal) to a spectrum analyser and to a power meter. A separate measurement made by replacing the Gunn-effect resonator with a short circuit gave the reflection coefficient of the resonator input for the $f_t$ values of interest. Additionally, the klystron signal power behind the attenuator was known. It was then possible to obtain the value of $P_i$ delivered to the Gunn-effect oscillator. The minimum bias voltage $V_{Bm}$ for $P_i$ just to succeed in nucleating oscillations as a function of $\Delta f = f_0 - f_t$ is shown in Figure 6.2 for two values of $P_i$ where $P_0$ is the output power produced by the Gunn-effect diode, whose $f_0$ was 8.49 GHz.

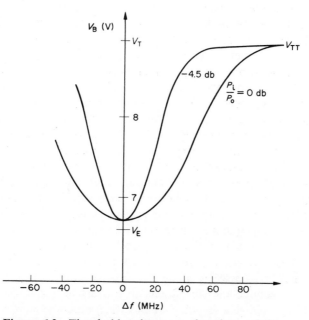

**Figure 6.2** Threshold voltages as functions of frequency shift $\Delta f$ and two values of input/output power ratios $P_i/P_0$ ($V_T$ is the threshold voltage for oscillation triggering by bias voltage alone; $V_{TT}$ is the threshold voltage for the simultaneous application of $P_i$, the frequency of which is different from that of $P_0$ by the value $\Delta f$; and $V_E$ is the extinction voltage).

Once oscillations have been triggered by $P_i$ and $P_i$ has terminated, the bias voltage has to be lowered below a certain extinction voltage $V_E$ in order to terminate oscillations. $V_E$ has the value of 6.75 V. The negative step-recovery module spike has therefore to be of sufficient amplitude to bring the bias voltage below $V_E$.

The threshold bias voltage for the case of $P_i = 0$ W for the mode under consideration is given by $V_T = 8.8$ V. For large values $\Delta f$, $V_{Bm}$ approaches $V_T$.

One can now give the criteria for successful operation of the regenerator. For the frequency shift $\Delta f = 0$, relating to the 'mark' frequency $f_1$, the bias voltage $V_B$ has to be above the line $V_E$ but below $V_{Bm}$. The positive retiming spike from the step-recovery module has to be of such an amplitude $V_{S+}$ that $V_B + V_{S+}$ is above $V_{Bm}$ at $\Delta f = 0$, but below $V_{Bm}$ at $\Delta f$ corresponding to the 'space' frequency $f_2$. The 'mark'-signal only then succeeds in triggering oscillations at the exact time given by the retiming pulse. The negative spike amplitude $V_{S-}$ from the step-recovery module must bring the difference between $V_B$ and $V_{S-}$ to below $V_E$ so that reliable oscillation extinction can be obtained.

F.S.K. pulse regeneration was obtained for 20 megabit/second operation as a first feasibility study, and evidence was found that operation at gigabit/second flow-rates should be possible. The output power at $f_0$ for the Gunn-effect oscillator employed was about 0.8 mW.

It is possible to find mode-switching effects which might, for example, be utilized for memory operation. In particular, the following phenomenon was observed for some diodes. When oscillations were started by raising $V_B$ above $V_T$, the amplitude was very much smaller than for the trigger case when $V_B$ is only above $V_{Bm}$ and $P_i$ is applied. The larger oscillation amplitude was maintained even after $P_i$ was reduced to zero. Obviously, $P_i$ raised the combination of diode and coaxial resonator into a higher-power mode of operation which could not be reached by the Gunn-effect diode oscillation alone. This behaviour was clarified by results on the negative device conductance $g_d$ vs. voltage swing $v_{ac}$ as obtained by the transient method described in section 2.4.2, page 41. For some diodes, characteristics such as those of Figure 6.3, page 121, were measured. For low bias voltages a single peak of $g_d(v_{ac})$ occurs, and the operating point would be somewhere along the negative slope of $|g_d(v_{ac})|$ as given by the load conductance in parallel to this active device. For a higher value of bias voltage a second peak of $g_d(v_{ac})$ occurs (Figure 6.3). This is caused by a different space-charge wave mode inside the diode, as substantiated by additional measurements on the diode-current behaviour. In fact, the first peak seems to be representative of a classical domain mode whereas the second one might be based on some hybrid mode. When a suitable load conductance is available, the operating point P′ is reached by applying the bias voltage, whereas P″ can only be reached by an external locking signal which lifts $v_{ac}$ across the second peak (Figure 6.3). Once P″ has been reached, it will of course be a stable condition, in agreement with experimental findings.[33]

It is suggested that these and other threshold effects of Gunn-oscillator modes can be utilized for applications such as advanced communication systems.

In order to see whether base-band diodes can trigger and amplitude-modulate $X$-band Gunn-effect oscillators, the following experiments were

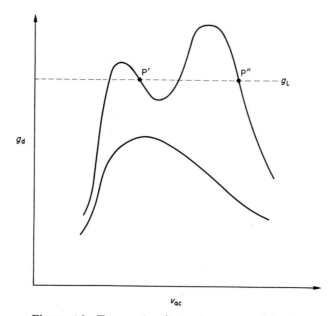

**Figure 6.3** Two-mode effects demonstrated by two peaks of the negative device conductance $g_d$ *vs.* a.c. voltage amplitude $v_{ac}$ across device at resonant frequency. For a given load conductance $g_1$ two stable operating points P′ and P″ are obtained, where P″ can only be obtained by injecting a large external $v_{ac}$ for a short time.

performed.[12] A base-band diode with an inter-electrode distance of $l = 155$ μm, a resistivity of 1.806 Ω cm, and a low-field resistance of 500 Ω, was operated under pulsed bias conditions. The bias pulse of 100 Hz repetition frequency and 220 ns duration was first applied to an integrating circuit so that the bias pulse frequencies were much lower than the domain-signal frequencies. The bias voltage was set at such an amplitude that a succession of domain pulses occurred at the peak value of applied bias. The Gunn-effect diode was earthed at the anode end. The bias voltage was applied via a 50 Ω coaxial cable, the impedance of which formed the series load resistance to the diode. The output pulses were then taken from across the diode and separated from the bias signals via a capacitive filter. The resulting Gunn-effect pulses of 1.5 V amplitude and of 750 megabit/second were applied to a low-$Q$, $X$-band coaxial Gunn-effect diode oscillator together with a d.c. bias voltage of about 5 V. The d.c. bias was adjusted in such a way that for a particular resonator-tuning-stub position the diode was operated near the threshold for microwave emission. A small increase in bias voltage then produces a large increase in microwave output power. The output was finally applied to the input of a sampling oscilloscope, which produced a signal trace representing the X-band microwave. The 10 GHz periods could not be displayed as this was outside the range of the sampling oscilloscope employed at that time. This trace showed very low, almost zero microwave power, except during production of the base-band pulses. As soon as the 750 megabit/second pulses were terminated, after about 80 ns, the microwave power returned to its original low level. The microwave pulse produced on the oscilloscope was about 13 mV in amplitude. This figure is admittedly low, but could be improved by further work, in particular with a better cavity design and matching.

It was established by a range of tests that the wave shape observed was in fact the base-band-modulated X-band output. Reducing the microwave-diode bias voltage by only a few per cent caused the 80 ns trace to disappear. Equally, an increase in this bias voltage produced high, continuous emission of microwaves and no pulse trace was observed any more. Detuning of the X-band cavity with its tuning screw also caused the pulse trace to vanish. In fact, the quality of performance was very sensitively dependent on tuning screw position. A reduction in base-band bias voltage of about 10 per cent eliminated any microwave pulse trace because no base-band domain signals were then produced, as seen on the sampling scope when fed with an inductive current probe monitoring the base-band diode current. The same effect was also observed when the earthing connection of the base-band diode was removed, so that no current was able to flow through the diode. Of course, this experiment did not yet show that an individual Gunn-effect pulse can trigger a few X-band periods. Separate measurements with very fast switching times of bias pulses have shown that subnanosecond pulse operation is in fact possible[39] by selecting diodes with $g_d(v_{ac})$ characteristics[41] which are suitable for very fast oscillation growth and decay.

To demodulate the F.S.K. signals they are applied to a band pass filter, permitting only $f_1$ to pass. This is given to a base-band regenerator Gunn-effect diode together with a suitable bias voltage.

The realization of Gunn-effect circuits for other modulation systems has to be considered. In particular, it should be possible to set up modulators, demodulators etc., for both P.S.K. and A.S.K. as well as additional elements for a F.S.K. system to those outlined above. Techniques such as phase-comparison phase-reversal keying with differential encoding can be employed also with Gunn-effect diodes. Phase-shift keying is possible where two microwave diodes with low-$Q$ loading are suitably inserted into a ring modulator. If one diode is oscillating during, say, the 'space' position, the other diode is switched off. This can be achieved with correct bias voltages, low-power microwave locking, and signals from base-band Gunn-effect diodes with suitable polarity so that one microwave diode receives a positive pulse when the other experiences a pulse of opposite polarity. The microwave diodes are positioned in such a way that the signals from one diode have to travel the time of 180° of the microwave phase longer than the other one. This gives a phase-modulated output at the output terminal of the circuit. It should also be possible to demodulate the P.S.K. signals by employing a microwave Gunn-effect oscillator which is frequency- but not phase-locked to the P.S.K. signals. The P.S.K. signals are also applied to a phase-locking Gunn-effect diode, acting as amplifier and limiter, and are then additively combined with the output from the frequency-locked Gunn-effect oscillator. The resulting envelope, obtained with a fast Schottky-barrier diode, represents the demodulated signal.

Further possibilities can be considered in order to fully exploit the new devices now available.

# 7: Further Logic Applications with Electric-field Domains

Gunn-effect domains can be controlled by various other means which have not yet been considered, and it is important to discuss them here although these possibilities are unlikely to result in any immediate commercial exploitation. Additionally, there are further high-field domains which do not depend on the transferred-electron effect. These are usually much slower than Gunn-effect domains and are therefore less attractive for device developments. However, it is felt necessary to review these here too.

## 7.1 DOMAIN SHAPING

It is possible to affect the size of the domain during its transit. If we take the induced current at the anode during transit, that is, approximately the current decrease due to the existence of a domain, we obtain

$$\Delta I_D = AJ \cong A\varepsilon \frac{dE}{dt}$$

With Poisson's equation for the domain field,

$$\frac{en_0(x)}{\varepsilon} = \frac{dE}{dx}$$

($n_0(x)$ is the ionized donor density at the point along the domain trajectory, $x$, where the domain is passing), and with the approximation

$$\frac{dE}{dt} = \frac{\partial E}{\partial x}\frac{\partial x}{\partial t} \cong \frac{\partial E}{\partial x} v(x)$$

one finds

$$\Delta I_D = ev(x_1)n_0(x_1)A(x_1) \tag{7.1}$$

where $x_1$ is the instantaneous position of the domain, $v$, $n_0$, and $A$ are the drift velocity, ionized donor density, and diode cross-sectional surface respectively for the domain at $x_1$. A change of any of these three terms causes $I$ to alter. This result is based on the effect that a fully developed domain can be understood to act as a wall of moving charge which controls the current at the contacts. The shape of $\Delta I_D$ can therefore be affected by any of the following points:

1. The variation of doping by impurity diffusion, epitaxy growth techniques, or similar means.

2. Changes in $A$. As the interelectrode distance $l$ has to be very short for the pulse rates of interest, photolithography has to be employed with etch techniques applied to epitaxial layers deposited on semi-insulating substrates.

3. Localized ionization of donors by high domain fields. When a first domain is in transit, the bias voltage is increased sufficiently at various instances so that field ionization occurs at various points along $l$. Subsequent domains then experience there an increased $n_0$ and their corresponding current waves are a replica of the voltage waveform applied when the first domain existed. This effect lasts as long as the life-time of excess carriers, and this can be up to a microsecond for favourable conditions. One has here a possibility of storing very fast wave functions. The limit is given by a Debye length.

4. Illumination of the $n$-layer can result in optical generation of excess carriers along $l$. Domains travelling along the resulting conductivity profile then produce a current waveform at the contacts which is a representation of the light pattern.

5. Placing some conducting layers or contacts along the surface short-circuits parts of the domain fields so that the effective domain area $A$ is reduced. For example, a $p^+n$-junction can be deposited on the semiconducting GaAs. This is kept reverse-biased so that the conducting $p^+$ layer has no influence on the domain. However, when local illumination ionizes the junction, a conducting path is formed and the resulting domain current again contains an indication of the light pattern. Instead of light, other ionizing methods are of course also possible, such as X-rays or nuclear irradiation. Another example is the deposition of some photoconductive semiconducting material such as CdS along the free surface of the $n$-layer.

6. Dielectric, magnetic, and resistive surface-loading materials can either affect the domain instability as described in Chapter 3, or they can produce strong space-charge layers. For example, highly dielectric materials covering only parts of the free surface of a Gunn-effect device can induce an accumulation layer near the anode and a depletion layer near the cathode, because the electrode potential is applied to the capacitance formed by the loading dielectric which is then charged up by arranging charge carriers in the semiconductor. The result is that domains are affected because effectively the carrier density available for depletion in the domain depletion layer is altered, and $\Delta I_D$ of equation (7.1) gives a waveform which is a representation of this effect.

When these domain-influencing parameters of equation (7.1) are drastically changed, domain extinction and also new domain formation can occur. In fact, logic systems have been proposed where very long GaAs structures with numerous domain extinction and nucleation electrodes (both ohmic and dielectric) are set up such that logic functions can be achieved. It is possible to envisage total logic systems such as a parallel adder for ultrafast logic applications. Such structures have to be set up in monolithic forms, and require high-quality layers of extended lengths. Indeed, after a first realization of Gunn-effect logic with the more simple elements of the previous chapters, this might represent the approach which ensures optimum exploitation of the fast speed possibilities of Gunn-effect domain logic.

Another application of the techniques described above is the production of wave function generators. By shaping the diode so that a given variation of $A$ is set up, for example, and wave function can be obtained. By having one part of the $n$-layer path with low electric bias-fields due to a local widening of $A$, the length of the domain trajectory can be controlled easily by a small change of bias voltage. For a lower bias level the domain will not succeed in crossing the part with increased $A$, whereas for a slightly increased bias voltage the domain passes this obstacle and can then run on in a subsequent further high-field part of the diode. Switching between two or more pulse repetition frequencies is thus easily possible.

By tapering the diode so that $A$ increases continuously towards the anode, the length of the domain trajectory can be made to be proportional to the applied bias voltage. The resulting pulse repetition frequency is therefore an indication of the bias voltage. If the $n$-layer is strongly indented in a periodic manner, the domain current fluctuates additionally with the period of the variation of $A$. This periodic signal fluctuation can be extracted either by passing the device current through a series load resistor, or by detecting the passage of a domain by a capacitive electrode which couples only to the peaks of the wavy surface. Then tapering and periodic variations in $A$ together result in a device which produces output pulse signals the number of which is proportional to the bias voltage. If this bias voltage is an analogue signal, the resulting output is the pulse-number representation of the analogue value. The output would have to be suitably coded to obtain analogue-to-digital conversion. Such device possibilities, which have been investigated experimentally by various laboratories, have been termed DOFIC, representing *d*omain-oriented *f*unctional *i*ntegrated *c*ircuits.[50]

## 7.2 OPTO-ELECTRONIC APPLICATIONS

There have been several optical effects which seem to make logic systems at optical frequencies an attractive possibility for the future. For example, GaAs junction laser arrays on a common substrate can be triggered to emit in certain directions depending on bias voltages applied, and this

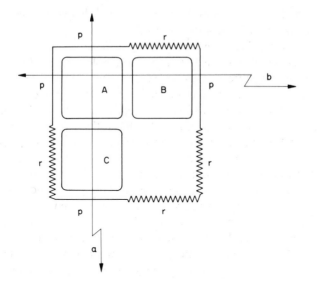

**Figure 7.1** Optical memory consisting of a monolithic structure with one substrate and three separate junction lasers A, B, and C on the same plane so that strong laser interaction between each junction is possible. The structure is produced by mesa-etching; the surfaces p are optically flat by cleavage or polishing, whereas the surfaces r are roughened.

performance can be employed to develop various logic gates. As an example, Figure 7.1 shows three junctions A, B, and C, where only the lateral surfaces p are polished, whereas those denoted by r are roughened. A roughened surface can inhibit lasing, because no optical resonance is set up. A current through junction A brings this device near the threshold for lasing. When a high current is temporarily applied to C, lasing is started along direction a only. This effect continues even after the current through C has been terminated. When a current pulse is applied to junction B, the lasing along direction a stops and emission is now set up along b. This again is continuous until a pulse is applied to C. The principles of the structure of Figure 7.1 have been employed in order to set up various logic gates and an adder. The disadvantage is that logic functions have to be obtained by applying high current pulses to junction devices. Therefore there is no advantage here yet with respect to conventional junction logic, except if special optical applications are called for.

However Gunn-effect domains, with their fast current-switching characteristics, may represent the solution. In fact the current pulses required for room-temperature, CW heterojunction lasers can be provided by Gunn-effect diodes, and a useful circuit would be a parallel connection of a Gunn-effect diode with a laser junction.[26] As soon as a domain is nucleated, additional current is then forced through the laser diode resulting in lasing during domain transit.

The disadvantage here is again the laser junction capacitance. The discovery that the high fields of a domain can be employed directly to produce laser action[54] is therefore of great interest. One hopes that a coherent optical pulse is emitted for each domain transit. These pulses would then be very sharp and short, and extremely high pulse rates can be expected as long as the domain fields are kept sufficiently high. In fact there is evidence with Southgate's results that a generated light pulse has about the same duration as a single domain transit. It is probable that this is due to the strong recombination effect occurring with stimulated emission of lasers. Additionally, the field-enhanced trapping of upper-valley electrons[59,61] might aid the fast removal of ionized electrons.

The reverse operation of nucleating domains when an optical pulse arrives can be obtained by employing a photosensitive load to the Gunn-effect diode. Some photoconductive semiconductors show very large resistivity ratios. In fact, in order to obtain well-defined maximum output signal pulses the Gunn-effect diode is loaded by a photoconductive and an additional ohmic resistance in series (see Figure 7.2). An example for

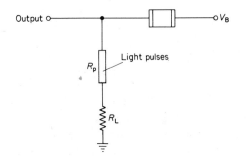

**Figure 7.2** Optical-to-electronic pulse transformer ($R_p$ photoresistor).

the resistance values is as follows: photoresistor dark, $R_{pd} = 10\ \Omega$; photoresistor light, $R_{pl} = 1\ \Omega$; Gunn-effect diode low-field resistance, $R_0 = 100\ \Omega$; and load resistance, $R_L = 20\ \Omega$. With $R_L$ and $R_{pd}$ the voltage drop across the diode has to remain below threshold, whereas under illumination the diode field is raised above threshold, as schematically shown by Figure 7.3. The output pulse $V_0$, as can be seen from Figure 7.3, is approximately independent of the state of the photoresistor if $R_L$ is larger than $R_{pd}$. The output pulse can then be considered a reshaped pulse signal, as the domain can continue its transit even if the optical pulse has terminated beforehand. A disadvantage at present is the long life-times of the optically-produced excess carriers. New techniques have to be developed for killing life-times, either by enhancing surface recombination or by introducing further bulk recombination centres.

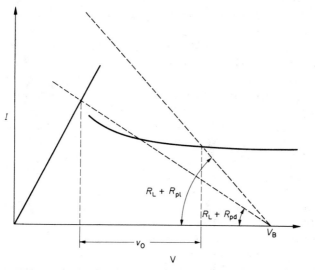

**Figure 7.3** Current-voltage characteristic of Gunn-effect diode with load lines for the illuminated photo-resistor and for the dark case ($v_0$ is the output voltage amplitude).

Direct illumination of the semiconducting GaAs layer near the anode has also been used successfully in order to raise the field near the cathode above the domain-nucleation threshold. Again recombination times limit the speed of operation, although by careful adjustment nanosecond recovery times have been achieved. Of interest here is the suggestion, which was advanced on theoretical grounds but which seems to have some experimental evidence, that there is a greater probability for recombination to occur from the satellite valley.[43] A passing domain could therefore also serve for wiping away excess carriers.

## 7.3 THE ACOUSTO-ELECTRIC DOMAIN

High-field travelling domains have been observed also in connection with acousto-electric instabilities as found in piezo-electric semiconductors. When, due to the application of a sufficiently high bias voltage, carriers drift faster than the acoustic waves in such piezoelectric crystals, the acoustic wave acquires power from the drifting electrons in a manner similar to that in which the travelling-wave tube has a transfer of energy from the electron beam to the electromagnetic wave along the slow-wave structure. This acousto-electric phenomenon, which has therefore a well-defined threshold again, produces a reduction in current for increased bias voltage. This is again a negative resistance effect. If the instability is strong enough, clearly defined travelling high-field domains occur which are similar to Gunn-effect domains. They are, of course, associated with strong acoustic wave-packets. They nucleate near the cathode electrode and travel to the anode with a velocity which is almost equal the acoustic velocity of $10^6$ cm/s. As soon as a domain reaches the anode it is extinguished, and a new one is formed near the cathode. A domain causes current saturation during its transit. Typical domain fields are more than $3 \times 10^4$ V/cm, and the domain width is around 150 μm. Acousto-electric domains have been reviewed previously and the reader is referred to pages 120–122 and 170–171 of reference 20.

It is possible to achieve numerous logic functions with acousto-electric domain devices, as with the Gunn-effect domain. The disadvantage here is the slower speed of acousto-electric domains and thus the reduced pulse rates achievable. However, there are a few applications where acousto-electric effects might be of advantage.

An acousto-electric light scanner has been developed, which is called SALS[19] (*s*olid-state *a*cousto-electric *l*ight *s*canner). A *n*CdS strip 400 μm wide, less than 100 μ thick and 0.2–0.4 cm long, with a resistivity of 1 Ω cm, has a *p*Cu$_2$S layer deposited on one surface. This layer is cut by photolithography into perpendicular strips. Ohmic contacts are applied on opposite ends of the CdS layer, and a travelling acousto-electric domain is formed by using a sufficiently large voltage between these contacts. On its trajectory the domain passes underneath the perpendicular *p*Cu$_2$S strips which form a *pn*-junction with the *n*CdS. The domain voltage causes a local breakdown in the junction and some current can bypass the domain, increasing the output current momentarily. The flow of current through the junction is accompanied by an emission of red light. A momentary increase in device current can produce an increased current through the junction with enhanced light output. It is therefore possible to produce an electronically controlled light-emission pattern. On the other hand, by illuminating the *p*Cu$_2$S layer and thus increasing its conductivity, the amount of current short-circuiting the domain can be increased. This is then seen as an increase in terminal current. The device can therefore also be used as a scanning light detector.

The acousto-electric domain travels about 100 times slower than the Gunn-effect domain. If one produces a sandwich of CdS, insulator, and *n*GaAs, acousto-electric domains can be made to travel in one direction, whereas Gunn-effect domains can be made to run perpendicular to this. By this technique, a point where both domains overlap is scanned across the whole sandwich in the same way as an electron beam is scanned across the screen of a television tube. This scanner has also only two input connections, one to each of the two anodes. A light beam can be modulated by a passing domain using Pockel's effect. In this way a laser beam could possibly be scanned before being projected onto a screen. The laser light source would, of course, have to be amplitude-modulated in the same way as the beam current of a television tube. It might thus be possible to develop a small optical system which can be employed to project films onto a screen. Owing to its small size and ruggedness, it might be superior to the present-day television set based on a fragile tube of relatively short life and of large size. However, substantial further work is required in order to achieve a working system. In particular, the quality of materials and their technology has still to be considerably improved.

GaAs is also a piezo-electric semiconductor. The Gunn-effect domain therefore has associated with it a bunch of acoustic waves, and this has been used by several experimenters (see e.g. reference 28) to generate bursts of acoustic waves at microwave frequencies. It might again be possible to influence the Gunn-effect domain shape or even to nucleate Gunn-effect domains by acousto-electric waves, although this field has not yet been fully explored.

# References

|  | Page |
|---|---|

1. ALBRECHT, P. 'Die Modulation von Gunn Oszillatoren bei Betrieb im Quenched-Domain Mode', *N.T.Z.* **24**, 1971, pp. 516–520.    41
2. BESTWICK, P., HARTNAGEL, H. L., HUNTINGTON, R. 'Frequency-shift Keyed Pulse Retiming with Gunn Oscillator', *Electronic Letters*, **6**, 1970, pp. 46–47.    117
3. BOHN, P. P., HERSKOWITZ, G. J. 'Impact Ionization in Bulk GaAs High Field Domains', *I.E.E.E. Trans. El. Dev.*, **ED-19**, 1972, pp. 14–21.    47, 49
4. BRUCE-CARLSON, A. *Communication Systems—an Introduction to Signals and Noise in Communication*, McGraw-Hill, 1968.    113
5. BUTCHER, P. N., FAWCETT, W., OGG, N. R. 'Effect of field-dependent diffusion on stable domain propagation in the Gunn effect', *Brit. J. Appl. Physics*, **18**, 1967, pp. 755–759.    12
6. CARROLL, J. E. *Hot Electron Microwave Generators*, E. Arnold, London, 1970.    3, 5, 41
7. COX, R. H., STRACK, H. 'Ohmic Contacts for GaAs Devices', *Solid State Electronics*, **10**, 1967, pp. 1213–1218.    85
8. DWIGHT, H. B. *Tables of Integrals and other Mathematical Data*, Macmillan, New York, 1964, p. 38.    26
9. EDWARDS, W. D., HARTMAN, W. A., TORRENS, A. B. 'Specific Contact Resistance of Ohmic Contacts to GaAs', *Solid State Electronics*, **15**, 1972, pp. 382–392.    85
9(a). ENGELMANN, R. W. H., HEINLE, W. 'Effect of Diffusion on the Small-signal stability of Semiconductor Plates with Negative A.C. Mobility', *Archiv für Electronik und Übertragungstechnik*, **27**, 1973.    57, 59, 64
10. FALLMANN, W., HARTNAGEL, H. 'Metallic Channels formed by High Surface Fields on GaAs Planar Devices', *Electronics Letters*, **7**, 1971, pp. 692–693.    88
11. FALLMANN, W., HARTNAGEL, H. L., MATHUR, P. C. 'Experiments on Heat Sinking of Semiconductor Devices', *Electronics Letters*, **7**, 1971, pp. 512–513.    91, 95
12. FALLMANN, W. F., HARTNAGEL, H. L., SRIVASTAVA, G. P. '*X*-Band Gunn Oscillators Triggered by Base-Band Gunn Diodes', *Electronics Letters*, **6**, 1970, p. 100.    121
13. FALLMANN, W., HARTNAGEL, H. 'Aspects of Planar Gunn Diodes for High C.W. Output Powers', *Solid State Electronics*, **14**, 1971, pp. 909–912.    90
14. FALLMANN, W. F., HARTNAGEL, H., MATHUR, P. C. 'Gunn Effect Logic Circuits', Progress Report 1.3.71. to 31.8.71, Dept. Electrical and Electronic Engineering, University of Newcastle upon Tyne, Contract NRDC 4298–02.    71
15. FISHER, R. E. 'Generation of Subnanosecond Pulses with Bulk GaAs, *Proc. I.E.E.E.* **55**, 1967, p. 2189.    115
16. GLENN, A. B. 'Comparison of PSK vs. FSK and PSK-AM vs. FSK-AM Binary Coded Transmission Systems', *I.R.E. Trans. Comm. Syst.*, **CS8**, 1960, pp. 87–100.    116
17. GUETIN, P. 'Contribution to the Experimental Study of the Gunn

Effect in Long GaAs Samples', *I.E.E.E. Trans. El. Dev.* **ED-14**, 1967, pp. 552–562. 67

18. HAISTY, R. W., HOYT, P. L. 'Investigation of Voltage Breakdown in Semi-insulating GaAs', *Solid State Electronics*, **10**, 1967, pp. 795–800. 95

19. HAKKI, B. W. 'Solid-State Acoustoelectric Light Scanner,' *Appl. Phys. Letter*, **11**, 1967, p. 153. 130

20. HARTNAGEL, H. *Semiconductor Plasma Instabilities*, Heinemann, London, 1969. 3, 5, 41, 130

21. HARTNAGEL, H. 'Theory of Gunn-Effect Logic', *Solid State Electronics*, **12**, 1969, pp. 19–30. 72

22. HARTNAGEL, H. 'Three-Level Gunn Effect Logic', *Solid State Electronics*, **14**, 1971, pp. 439–444. 73

23. HARTNAGEL, H. 'Three-terminal Gunn Logic', *Archiv. d. El. Übertragg*, **23**, 1969, p. 527. 78

24. HARTNAGEL, H. 'Reflection Insensitive Gunn Regenerator for Pulse Communication', *Solid State Electronics*, **14**, 1971, pp. 1331–1333. 78

25. HARTNAGEL, H., IZADPANAH, S. H. 'High-speed computer logic with Gunn-effect devices', *The Radio and Electronic Engineer*, **36**, 1968, pp. 247–255. 108

26. HARTNAGEL, H. L. 'Pulse Communication using Gunn Diodes and Heterojunction Lasers', *Archiv für Elektronik und Übertragungstechnik*, **25**, 1971, p. 51. 127

27. HARTNAGEL, H., HUTSON, V. C. L. 'Thermal Resistance of Planar Semiconductor Structures', *Proc. I.E.E.* **119**, No. 6, 1972, pp. 655–658. 93

28. HAYAKAWA, H., ISHIGURO, T. TAKADA, S., MIKOSHIBA, N., KIKUCHI, M. 'Generation of High-Frequency Ultrasonic Waves by Gunn Effect', *J. Appl. Phys.* **41**, 1970, pp. 4755–4762. 131

29. HOENEISEN, B., MEAD, C. A., 'Power Schottky Diode Design and Comparison with the Junction Diode', *Solid State Electronics*, **14**, 1971, pp. 1225–1236. 3

30. HOFMANN, K. R., 't Lam, H. 'Suppression of Gunn Domain Oscillations in thin GaAs Diodes with Dielectric Surface Loading', *Electronics Letters*, **8**, 1972, pp. 122–124. 57

31. HOFMANN, K. R., 'Stability Theory for thin Gunn Diodes with Dielectric Surface Loading', *Electronics Letters*, **8**, 1972, pp. 124–125. 57

32. HOLLAND, M. G., 'Thermal Conductivity', *Semiconductors and Semimetals*, **2**, 1966, p. 3. Editors: P. K. Willardson and A. C. Beer, Academic Press, New York. 91

33. HUNTINGTON, R., HARTNAGEL, H. O., BESTWICK, P. 'Signal Injection Triggering of X-Band Gunn Oscillators for F.S.K. Operation', *Electronics Letters*, **6**, 1970, p. 234. 120

34. IKOMA, T., SUGETA, T., TORIZUKA, H. YANAI, H. 'Characteristics of the Transferred Electron Devices', *J. of Faculty of Engineering, University of Tokyo*, **30**, 1970, pp. 347–394. 33

35. IZADPANAH, S. H., HARTNAGEL, H. L. 'Gunn-effect Pulse and Logic Devices', *The Radio and Electronic Engineer*, **39**, 1970, pp. 329–339. 71

36. JUTZI, W. 'Ein Vergleich von MOS-und MES-Feldeffekttransistoren mit 1 μm Kanallänge fur integrierte Gleichstrom-gekoppelte Schaltungen', *Elektron. Rechenanl.* **14**, 1972, pp. 19–27. 110

37. JUTZI, W. 'Direct Coupled Circuits with Normally-off GaAs Mesfets at 4.2°', *Archiv. für Elektronik und Übertragungstechnik*, **25**, 1971, pp. 595–598. 110

38. KATAOKA, S., KOMAMIYA, K., MORISUE, M. 'A High-Speed Adder Using Gunn Diodes', *Proc. I.E.E.E.*, **59**, Oct. 1971, pp. 1526–7.          112

39. KAWASHIMA, M. HARTNAGEL H. L. 'A Fast Frequency-Switching System for FSK-PCM' *Electronics Letters* **7**, 1971 pp. 761–763.          122

40. KAWASHIMA M. KATAOKA, S. 'Measurement of Transverse Spreading Velocity of a High-Field Domain in a 3-Terminal Gunn Device', *Electronics Letters*, **6**, 1970, pp. 781–783.          78, 117

41. KAWASHIMA, M., HARTNAGEL, H. 'New Measurement Method of Gunn Diode Impedance', *Electronics Letters*, **8**, No. 12, 1972, pp. 305–306.          44, 122

42. KHANDELWAL, D. D., CURTICE, W. R. 'A Study of the Single-Frequency Quenched-Domain Mode Gunn-effect Oscillator', *I.E.E.E. Trans.*, **MTT-18**, 1970, pp. 178–187.          41, 47

43. KIMURA, T., YANAI, H., KAMIYAMA, M. (Dept. of Electronic Engineering, University of Tokyo, Japan): 'Optical Triggering of a Gunn Effect Device', Paper presented at Meeting of I.E.C.E. Technical Group on Semiconductors and Transistors July 1970, Paper No. SSD-70-16; and I.E.C.E. National Convention August 1970, Paper No. 726, p. 785 in Proceedings. (I.E.C.E.-Inst. of Electronics and Communication Engineering in Japan).          129

44. LAX, B., BUTTON, K. J. *Microwave Ferrites and Ferrimagnetics*, McGraw-Hill 1962, New York.          61

45. MAUSE, K. 'Simple Integrated Circuit with Gunn Devices', *Electronics Letters*, **8**, 1972, p. 62.          81

45(a). MAUSE, K., SALOW, H., SCHLACHETZKI, A., BACHERN, K. H., HEIME, K., 'Circuit Integration with Gate-controlled Gunn Devices', *Proc. 1972 Symposium on GaAs*, Boulder, pp. 275–285.          85

46. OWENS, R. P., CAWSEY, D. 'Microwave Equivalent-Circuit Parameters of Gunn-effect-device Packages', *I.E.E.E. Trans. Microwave Theory and Techniques*, **MTT-18**, 1970, pp. 790–798.          43

47. PAOLA, C. R. 'Metallic Contacts for GaAs', *Solid State Electronics*, **13**, 1970, pp. 1189–1197.          85

48. REISER, M., WOLF, P. 'Computer Study of Submicrometre fet's', *Electronics Letters*, **8**, 1972, p. 254.          110

49. RUCH, J. G., KINO, G. S. 'Transport Properties of GaAs', *Phys. Rev.*, **174**, 1969, pp. 921–931.          8

50. SANDBANK, C. P. 'Synthesis of Complex Electronic Functions by Solid-State Bulk Effects', *Solid State Electronics*, **10**, 1967, p. 369.          60, 126

51. SCHARFETTER, D. L., GUMMEL, H. K. 'Large-Signal Analysis of Silicon Read-Diode Oscillator', *I.E.E.E. Trans. Electron Devices*, **ED-16**, 1969, pp. 64–77.          47

52. SCHWARTZ, B. (Editor). 'Ohmic Contacts to Semiconductors', Electronics Division, The Electrochemical Society Inc. 1969, Conference Proceedings. (There are numerous important contributions).          85

53. SLATER, J. C. *Microwave Electronics*. Van Nostrand Co., 1950.          44

54. SOUTHGATE, P. D. 'Laser Action in Field-Ionized Bulk GaAs', *Appl. Phys. Letters*, **12**, 1968, p. 61.          128

55. SUGETA, T., IKOMA, T., YANAI, H. 'Bulk Neuristor Using Gunn Effect', *Proc. I.E.E.E.* **56**, 1968, pp. 239–240.          60, 64, 82

56. SUGETA, T., YANAI, H., SEKIDO, K. 'Schottky-gate bulk effect digital devices', *Proc. I.E.E.E.*, **59**, 1071, pp. 1629–1630.          82

57. SUGETA, T., YANAI, H. 'Logic and Memory Applications of the

Schottky-gate Gunn-effect Digital Device', *Proc. I.E.E.E.* **60**, 1972, pp. 238–240.                                                                                      100

58. TAKEUCHI, M., HIGASHISAKA, A., SEKIDO, K. 'GaAs Bulk Effect Pulse Regenerator with a Schottky Barrier Control Gate', *Proc. I.E.E.E.* **60**, 1972, p. 128.                                                      51, 82, 90, 95

59. TAKEUCHI, M., HIGASHISAKA, A. SEKIDO, K. 'GaAs Planar Gunn Diodes for DC-Biased Operation', *I.E.E.E. Trans. El. Devices*, **ED-19**, 1972, pp. 125–127.                                                              50, 128

60(a). THIM, H. 'Stability and Switching in Overcritically Doped Gunn Diodes', *Proc. I.E.E.E.* **59**, 1971, pp. 1285–1286.                           3

60(b). THIM, H. 'Experimental Verification of Bistable Switching with Gunn Diodes', *Electronics Letters*, **7**, 1971, pp. 246–247.                  3

61. TOKUMARU, Y., MIKOSHIBA, N. 'Bulk Negative Resistance enhanced by Trapping through upper Valleys in Semiconductors', *Proc. 2nd Japanese Conference Solid State Devices*, 1970. Supplement to *J. Jap. Soc. Appl. Phys.* **40**, 1971, pp. 107–113.                        50, 128

62. TOMIZAWA, K., KATAOKA, S. 'Dependence of Transverse Spreading Velocity of a High-Field Domain in a GaAs Bulk Element on the Bias Electric Field', *Electronics Letters*, **8**, 1972, pp. 130–131.           78, 111

63. ULLRICH, D. 'Observation of Recombination Radiation in Planar Gunn-effect Devices', *Electronics Letters*, **7**, 1971, p. 193.                 67

64. VILMS, J., GARRETT, J. P. 'The Growth and Properties of LPE GaAs', *Solid State Electronics*, **15**, 1972, pp. 443–455.                           85

65. WILLIAMS, R. 'Avalanche and Tunnelling Currents in GaAs', *R.C.A. Review*, **27**, 1967, pp. 336–340.                                              50

# Index